THE ORIGIN OF HUMAN
SOCIAL INSTITUTIONS

PROCEEDINGS OF THE BRITISH ACADEMY · 110

THE ORIGIN OF HUMAN SOCIAL INSTITUTIONS

Edited by
W. G. RUNCIMAN

Published for THE BRITISH ACADEMY
by OXFORD UNIVERSITY PRESS

Oxford University Press, Great Clarendon Street, Oxford OX2 6DP

Oxford   New York

*Athens   Auckland   Bangkok   Bogotá   Buenos Aires   Cape Town
Chennai   Dar es Salaam   Delhi   Florence   Hong Kong   Istanbul   Karachi
Kolkata   Kuala Lumpur   Madrid   Melbourne   Mexico City   Mumbai   Nairobi
Paris   São Paulo   Shanghai   Singapore   Taipei   Tokyo   Toronto   Warsaw*

*and associated companies in
Berlin   Ibadan*

*Published in the United States by
Oxford University Press Inc., New York*

*British Library Cataloguing in Publication Data
Data available*

*ISBN 0–19–726250–3
ISSN 0068–1202*

Typeset by J&L Composition Ltd, Filey, North Yorkshire

*Printed in Great Britain
on acid-free paper by
Bookcraft Ltd
Midsomer Norton, Somerset*

Contents

Notes on Contributors

Ofer Bar-Yosef is Professor in the Department of Anthropology, Peabody Museum, Harvard University.

Jerome H. Barkow is Professor in the Department of Sociology and Social Anthropology, Dalhousie University, Halifax N.S.

Ken Binmore is Professor of Economics at University College London, Director of the ESRC Centre for Economic Learning and Social Evolution, and a Fellow of the British Academy.

Robert Boyd is Professor of Anthropology at the University of California Los Angeles.

Richard Bradley is Professor of Archaeology at the University of Reading and a Fellow of the British Academy.

Robert Foley is Reader in Evolutionary Anthropology in the Department of Biological Anthropology at the University of Cambridge, and a Fellow of King's College, Cambridge.

Colin Renfrew is Disney Professor of Archaeology at the University of Cambridge, Director of the McDonald Institute for Archaeological Research, and a Fellow of the British Academy.

Peter J. Richerson is Professor in the Department of Environmental Science and Policy at the University of California Davis.

W. G. Runciman is Senior Research Fellow, Trinity College, Cambridge, and President of the British Academy.

Alasdair Whittle is Distinguished Research Professor in the School of History and Archaeology at Cardiff University, and a Fellow of the British Academy.

Preface

The first four papers in this volume were delivered at an open meeting held at the Academy's premises on 28 April 2000. The second four were delivered at a meeting limited to invited discussants which was held on the following day at the premises of the Novartis Foundation. My own concluding paper was added subsequently when I found that I had more to say than could be accommodated within an editorial introduction; it presents no original findings of any kind, but I hope that if nothing else it may help other sociologists to recognize the relevance to their own theoretical concerns of the issues raised in the volume as a whole.

The speakers and discussants were all chosen with a view to promoting discussion between archaeologists and other behavioural scientists sharing a common evolutionary perspective. My impression is that a number of them found it easier and more rewarding to communicate across disciplinary boundaries than they had initially expected. But at all events, the meetings confirmed the extent to which a thorough understanding of the critical early stages of human sociocultural evolution requires the collaboration of a range of disciplines, each of which has its own distinctive contribution to make.

My thanks are due to Rosemary Lambeth at the Academy and Gregory Bock at the Foundation for their support, to Hilary Edwards for secretarial and editorial assistance, and to James Rivington for his help and guidance in relation to publication.

W. G. R.
Trinity College, Cambridge

From Sedentary Foragers to Village Hierarchies: The Emergence of Social Institutions

OFER BAR-YOSEF

THE AIM AND SCOPE

IDENTIFYING THE TRACES OF VARIOUS TYPES of social organizations and institutions in the archaeological residues by reference to those known to us from historical and ethnographic records is notoriously not a simple task. It is a particularly precarious research endeavour when we try to decipher the material evidence of prehistoric populations of Late Palaeolithic and Early Neolithic age in south-west Asia, among which the first chiefdoms and states emerged in later times. The realization that the Neolithic Revolution resulted in a complex socioeconomic evolution, which followed variable trajectories in various regions of Eurasia and North Africa, has recently accelerated research into the timing and causes of the transition from the foraging mode of production to agriculture. The initiation of intentional cultivation, which eventually culminated in plant and animal domestication, meant major changes not only in workloads, division of labour, and permanent storage facilities, but essentially in the establishment of communal and private property, and increased control over territories. Hence, the new social structures of sedentary groups that replaced the egalitarian mobile foragers enjoyed rapid population growth and an increase in social inequality. In addition, these larger village communities were substantially more vulnerable to the impact of abrupt climatic changes.

Before delving into 'when', 'where', and 'why' this process took place in south-western Asia, we need briefly to review the theoretical aspects involved in the study of those societies, often referred to as 'small-scale', 'intermediate', or 'middle range' societies (e.g. Arnold 1996, and papers therein; Earle 1997; Johnson 1987; Price & Gebauer 1995, and papers therein; Upham 1990).

The entire sequence of socioeconomic changes, whether leading towards social complexity with ranked and stratified societies, or reversing its course

Proceedings of the British Academy, **110**, 1–38, © The British Academy 2001.

(a subject rarely discussed in the literature), was examined first by social anthropologists and more recently by archaeologists. The general notion is that the historical records, and by reference the archaeological evidence, document a global shift from egalitarian to non-egalitarian or complex societies that finally led to the emergence of chiefdoms and states. Therefore, in the context of this chapter, a short discussion of the implicit and/or explicit assumptions concerning the relationship between the primary sources of information such as the historical records and the secondary sources as derived from archaeological observations is needed. The recent advances emphasizing the role of human agency in the formation processes of social institutions will be mentioned, but the full treatment of the subject is beyond the scope of this chapter.

In the course of classifying the socioeconomic state of a given society and its place along the continuum from an egalitarian to a non-egalitarian society, most authors derive the terminology and models employed in their interpretations of the archaeology from the ethnographies of various world regions. Prominent among these are the North American, the New Guinean, and the Pacific Islands cases, as well as a few sub-Saharan African examples. Not surprisingly, while analysing the intricate social structure of non-egalitarian societies, whether related to power and labour (e.g. Arnold 1996; Earle 1997), or household sizes and social stability (e.g. Ames 1991, 1995; Coupland 1996), the authors rely on continuity between the regional ethno-historical information and the immediately preceding archaeological remains. However, in most, if not all of these cases, the transition from simple hunter-gatherers to non-egalitarian foragers is hardly explained, perhaps because the historical records, including oral traditions, do not provide ample evidence for the primordial time of a given group. We are therefore left with a quasi-static summary of the social components within non-egalitarian societies, be they of foragers or farmers, and the processes that led to increasing social complexity (e.g. Ames 1995; Hayden 1995; Johnson & Earle 1987; Price & Gebauer 1995).

From the presentations of the various basic models and the terminology for every step, phase, or grade of each organizational level, it becomes obvious that the social process and the region in which it took place could be homogeneous or heterogeneous. Hence, features of physical geography and the size of the region measured in square kilometres, as well as the nature of the resources, count. What distinguishes the Levant, as a sub-region of western Asia, is that within a short transect (e.g. 80–150 km) one finds an almost globally unmatched topographic and vegetational heterogeneity. This observation, for example, can be quantified by the number of plant species per km^2 (Danin 1988). The Levant has 0.0855 species per km^2, a little over twice the number found in coastal California, Greece, or countries in temperate Europe. Hence, as a region, the Levant differs in size, resources, and seasonality from those previously mentioned regions often employed by archaeologists as their main sources of information for societal evolution. In addition, the Levant is about one-third the size of the north-west coast of North America and about one-fifth that of New Guinea. In sum, comparing the three of them, major differences emerge in the number, kind, and distribution of potential food resources, whether vegetal or animal, and in their seasonal availability. Therefore, apart from hunting, acquisition techniques also differ.

In describing the archaeological data from the Levant and their social interpretation, like other modellers, I recognize that the common denominator for all societies is human nature. When the latter is detailed at the level of simple hunting and gathering societies this means competition among individuals, the presence of alpha-males, the acceptance of temporary or permanent coercion by all members of the group, and active participation in all realms of daily life that are classified as social interaction, technology, subsistence, and ritual (e.g. Service 1962). The second type of social organization to be employed in this chapter will be that of the 'non-egalitarian society', as defined through a series of attributes by Kelly (1995). This stage in societal evolution was previously called 'complex society' (Price & Brown 1985). The chapters in that volume already exhibited the variability within this class. As pristine non-egalitarian societies (and those — dare I say? — in a 'core area') are generally semi- to fully sedentary, the next phase would be the emergence of the headman, probably as a hereditary position. Spatially, at that stage we notice hierarchies among villages and hamlets in which local headmen wield power, prior to their full control by a chief and the formation of a chiefdom.

Applying social interpretation to archaeological remains is far from being an easy task. It is well known that the archaeological evidence, even when preservation is excellent (i.e. littered with organic remains), and recording by the excavators meets the standards of the day, is open to diverse interpretations. Therefore there is an obligation on the interpreter explicitly to enumerate the archaeological attributes that signify a given social state (for the society) and/or the presumed status of the individual. For this purpose, archaeologists tend primarily to exploit information from graveyards, the size and contents of houses, inter-village organization, and the nature and degree of long-distance exchange (or trade). A few examples are in place.

Among sedentary or semi-sedentary foragers, the presence or absence of decorated skeletons (with the distinctions of male/female and age at death) is taken as reflecting their status. In village societies, a gradation from simple burials to elaborated ones with rich body decorations and grave offerings symbolizes the individual's and/or household's status within the community. The lack of decorated skeletal relics in a village community is seen in a recent archaeological case as motivated by the need to negotiate equality among members (e.g. Kuijt 1996). Variability in the size of domestic buildings is considered as expressing the wealth of a family or household (Byrd 2000).

Driven by the need to interpret phenomena on a larger scale than that of a particular site, archaeologists exploit information gathered from several sites. In such an overview we are fully aware that we may expect differences between a 'core area' and the 'periphery'. This socioeconomic distinction, which can, but does not always, overlap with a sociopolitical classification, relates not only to world systems and industrial advances, but also to socially constructed concepts, ideologies, and material elements. The notion that 'core areas' and 'peripheries' existed in the prehistoric past was not uniformly accepted. Three decades ago, archaeologists saw all past hunter-gatherer groups across the continents as forming a social and biological continuum. Every subdivision proposed on the basis of lithic analysis was considered as an artificial splitting of past societies by researchers who wanted to identify 'archaeological cultures'. The 'splitters' were often blamed for 'creating prehistoric ethnicities' while constructing a cultural-historical sequence that was reflected exclusively in material elements.

Recently, the realization that genes, languages, and material culture do not necessarily co-vary facilitated a return to the basic aim of investigating archaeological cultures, which do not necessarily represent past ethnicities, and their changes through time. Such studies are aided by advances in the anthropology of technology and the improved understanding of the role of human agency in the past (e.g. Dobres & Hoffman 1994; Lemonnier 1992). None of these approaches to 'archaeology as history' pretends to identify 'ethnicity', language, evidence for gene drift, or the direction of gene flow. While the investigation of material culture may seem tedious and unrewarding research, the scientific advances resulting from it have enabled us to identify various important aspects of human behaviours such as subsistence systems (e.g. seasonality, patterns of hunting and butchering), stone tool making and usage (e.g. raw material procurement, core reduction techniques, curation of selected artefacts and function), building techniques, and the variable use of natural shelters, home ranges, and territories.

In the following pages, I will try to demonstrate how modes of production, fluid and/or conservative societal structures, and environmental fluctuations intertwine to form a reasonably coherent social history portraying how in south-west Asia complex hierarchical village communities emerged from semi-sedentary and sedentary groups of foragers.

THE CYCLICAL NATURE OF SEDENTISM

The term 'sedentism' is defined as the permanent presence of humans in a given place, which is often the interpretation given to a locale incorporating a few dwellings and sufficient evidence for more than seasonal occupation.

Habitations range from natural caves, rockshelters, and pit-huts to built-up houses of variable dimensions, among which palaces are ranked at the top of the scale. Sedentism is seen as the essential first step in the evolution of complex societies. Unfortunately, not all built-up environments were created by permanent settlers. There are known cases of mobile groups such as herders investing in the construction of shelters within the boundaries of their anticipated year-round routes.

Research by social anthropologists has demonstrated that the built environment of every society embodies the symbols for land ownership. Studies of basic architecture, kinship, and economic relationships within 'House Societies' (Lévi-Strauss 1983) were generally conducted in contemporary villages, or on cases known from historical contexts (e.g. Oliver 1971; Rapoport 1969). It is therefore not surprising that a cluster of houses, recognized as a hamlet or village, was often taken to represent a sedentary community (e.g. Rafferty 1985). However, given the ambiguities in the interpretation of the prehistoric remains, only houses in villages and more especially in towns from the time of the Bronze Age can be taken as evidence for year-round habitations. Prehistoric sites require a more careful approach. Following the pioneering work of Tchernov (1991a, 1993a), evidence for Late Pleistocene and Early Holocene sedentism is based on the presence of high frequencies of commensals ('self-domesticated species') among the bones of microvertebrates and birds, for example the house mouse, rat, and house sparrow. In semi-arid areas such as Sinai, the spiny mouse was found to be a commensal (Haim & Tchernov 1974). The archaeological attributes of sedentism, in descending order of confidence, are arranged in Figure 1.

All scholars agree that sedentism has both cultural and biological effects (Belfer-Cohen & Bar-Yosef 2000, and references therein; Rosenberg 1998, and references therein). However, investigators are divided on the issue of 'why' human groups became sedentary. Two alternative explanations are often mentioned. The first suggests that sedentism was caused by the attraction of humans to spatially restricted and rich resources where the 'law of least effort' would enhance the desire to stay for many months in the same locale. This process is known as the 'pull' model (e.g. Stark 1986). The second explanation proposes that economic and social circumstances enforced sedentism (Henry 1989; Keeley 1988; Kelly 1991; Rosenberg 1998, and references therein). The latter scenario is applicable, for example, to a situation in which abrupt climatic change and population densities in nearby territories impose reduced mobility, which results in social and technological changes. In both models, the decision to become sedentary seems to occur when a permanent camp allows for optimal exploitation of resources (both K- and r-selected). In addition, in both trajectories there is plenty of room for integrating the role of social concerns. For example, a rebellious decision by females and older members of the

Ofer Bar-Yosef

ARCHAEOLOGICAL MARKERS OF SEDENTISM
(or semi-sedentism)

BIOLOGICAL

THE PRESENCE OF COMMENSALS
KNOWN AS 'SELF-DOMESTICATED' SPECIES

Mus musculus domesticus – House mouse

Passer domesticus – House sparrow

Rattus rattus – Black rat

Acomys cahirinus – Spiny mouse (only in arid areas)

EVIDENCE FOR SEASONAL OCCUPATION

Carbonized plant remains (gathering and harvesting from February to November)

Gazelle cementum increments (summer/winter)

ARCHAEOLOGICAL

Pit-huts and houses

Permanent storage facilities

Heavy-duty tools such as mortars

Figure 1. The archaeological markers of sedentism, in descending order of importance.

group might force younger, more active men to stay nearby. Another issue is the nature and degree of 'population pressure'. Archaeologists often took this to mean population growth, and acknowledged that foragers, as far as is known today, regulated their populations in order to survive in a given environment. Alternatively, low reproductive rates are mentioned (Bentley *et al.* 1993). The archaeological testing of the physical evidence for population increase, especially in the Late Pleistocene period, is plagued with uncertainties that emanate from the degree of sites' visibility, recovery techniques, and the risk of gross errors in estimating the number of humans at a given time. However, various archaeologists observed a population growth during this time in the Levant (e.g. Bar-Yosef & Belfer-Cohen 1989b; Henry 1989; Moore 1989). Under such circumstances, when mobility must be reduced within a given territory, people would choose either to settle down or to expand and face physical conflicts with their neighbours. Similar resolution may be reached among territorially bounded people in a situation when 'population pressure' is a culturally construed, but economically unjustified, perception. Today, for example, social concerns are reflected in road signs in the United States that cite a place as being 'thickly settled' in areas that by Near Eastern standards would be considered as 'sparsely settled' (Bar-Yosef 1997b).

Sedentism has its price in sociopolitical, economic, and health tolls. Daily life in a village larger than a foragers' band heralds the restructuring of the social organization (Flannery 1972), as it imposes more limits on the individual as well as on entire households. To ensure the long-term predictability of habitable conditions in a village, members accept certain rules of conduct that include, among other things, the role of leaders or headmen (possibly the richest members of the community), active or passive participation in ceremonies (conducted publicly in an open space), and the like. The archaeological correlates for most of these aspects are commonly uncovered in Neolithic sites, as demonstrated by site reports and syntheses (e.g. Aurenche & Kozlowski 1999; Cauvin 1997; Kuijt 2000a, and papers therein; Voigt 1990).

The organizational resilience embedded in human societies, at least since the Upper Palaeolithic, demonstrates that among foragers sedentism may have been a temporary solution alternating with periods of mobility. The range of options within the system of residential and/or logistical mobility (Binford 1980) allowed, when conditions were right, the establishment of sedentary camps. It is suggested, for example, that the late Gravettian (24,000–20,000 BP uncalibrated) sites of Kostenki I and Avdeevo, where spatially organized pithuts have a superstructure of mammoth bones and equidistant layout of hearths, were sedentary camps (Serguin 1999). The seasonal information is interpreted as supporting year-round presence and is based on evidence of newborn mammoths (in springtime), dental analysis of polar fox (indicating late autumn/winter), and summer plant remains. In addition, the accumulations of mammoth bones in Gravettian sites in Moravia (Jelinek 1999) may reflect long-term occupations (Svoboda et al. 1996). Similar proposals were suggested for the Late Palaeolithic (after 18,000 BP) sites along the Dnestr River (Borziyak 1993).

It might be assumed that when additional seasonal information from Upper Palaeolithic and later sites is gathered or re-analysed we will recognize that human occupation was cyclical and that sedentism occurred in the past in more than one region or at more than one time. Therefore, the decisive transition from sedentary communities of foragers to those of cultivators becomes an import juncture, a 'point of no return' causing an avalanche in the history of human settlements. This process, which must have been unique in every region, must be investigated before we can focus on the ensuing sociocultural changes. It seems that in the Near East, there was a close relationship between the worsening climatic conditions (which determine the geographic distribution of annual and permanent food resources) and the onset of cultivation by sedentary foragers.

CLIMATIC IMPACTS ON LATE PLEISTOCENE–EARLY
HOLOCENE PAST SOCIETIES

In studying human responses to natural disasters such as droughts, floods, and sea-level rise — which are often expressed in famines, migrations out of the affected region, the collapse of communities, and death — researchers in modern relief programmes have found that each population negotiates the impacts of natural disaster in its own way. While physically abrupt climatic calamities may be similar, human reactions differ. In classifying the components of a safeguard or buffer mechanism against the effects of prolonged stress or an abrupt disaster of a given population, investigators identify the factors that would enable the individual and the group to cope and to reach an operative decision. The cultural components depend on the group's size and its social organisation (i.e. band, macro-band, tribe, or state), as well as on the size of the affected region (Glantz 1987, 1994; Glantz *et al.* 1998). In brief, in evaluating the impact of natural disasters, whether of long or short duration, we need to take into account the particular 'cultural filter' of the affected population. The application of this cumulative knowledge to the particular conditions of south-western Asia is already summarized in various forms in the literature (e.g. Bar-Yosef 1996; Bar-Yosef & Belfer-Cohen 1992; Goring-Morris & Belfer-Cohen 1997; Henry 1997; Moore & Hillman 1992; Sanlaville 1997).

The current review of the Terminal Pleistocene through the first half of the Holocene is based on calibrated BP dates calculated through INTCAL98 or Calib.4.2 (Stuiver *et al.* 1998). It incorporates the main archaeological entities of the Late Epi-Palaeolithic and the Early Neolithic (Figure 2). In order to keep the overall chronology in a simple form, a choice was made by the author. Most radiocarbon dates have more than one optional calibrated date, especially if calculated within two SD (95 per cent of probability). However, I have used the median figure produced by the software, as calculated with only one standard deviation. Obviously this procedure does not mean that the calibrated date is calendrically accurate. Despite these limitations, the proposed procedure facilitates testing the degree of correlation between the archaeological manifestations and the climatic fluctuations.

In recent years, the Greenland ice cores (GRIP and GISP) have become the main sources for proxy palaeoclimatic conditions during the Terminal Pleistocene and Early Holocene in the northern hemisphere (e.g. Alley *et al.* 1993; Mayewski & Bender 1995; O'Brien *et al.* 1995). Similar information is gathered from marine and terrestrial sources, and the degree of synchronization between all these data sets is repeatedly tested. This type of investigation uncovered a major caveat. In the eastern Mediterranean, dated pollen records from marine cores demonstrated a lack of synchronization with pollen sequences from lake cores in the order of hundreds or even thousands of years

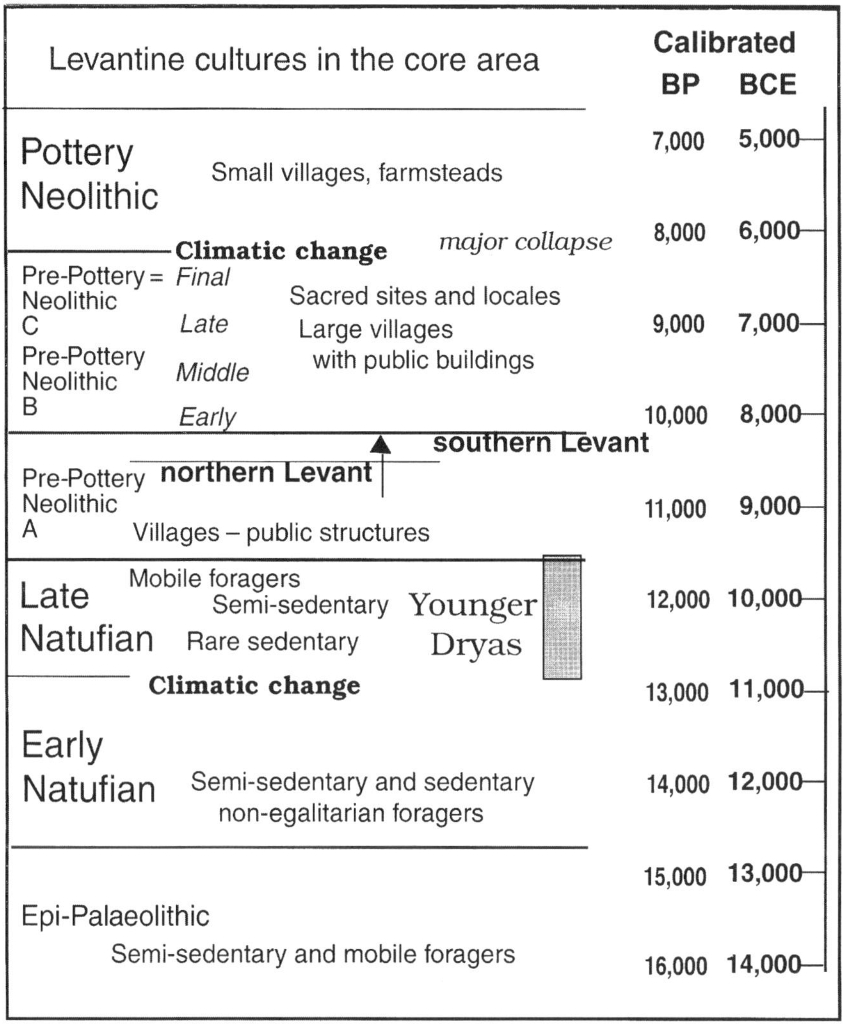

Levantine cultures in the core area	Calibrated BP	BCE

Pottery
Neolithic   Small villages, farmsteads   7,000   5,000

——————————— Climatic change   *major collapse*   8,000   6,000

Pre-Pottery = *Final*
Neolithic   Sacred sites and locales
C   *Late*   Large villages   9,000   7,000
Pre-Pottery   with public buildings
Neolithic   *Middle*
B
                 Early   10,000   8,000
                                    ▲ **southern Levant**
Pre-Pottery **northern Levant**
Neolithic   11,000   9,000
A   Villages – public structures

Late   Mobile foragers
                    Semi-sedentary   Younger   12,000   10,000
Natufian   Rare sedentary   Dryas

——————————— Climatic change   13,000   11,000

Early
Natufian   Semi-sedentary and sedentary   14,000   12,000
                non-egalitarian foragers

                                                                15,000   13,000

Epi-Palaeolithic
        Semi-sedentary and mobile foragers   16,000   14,000

Figure 2. Chronological chart of calibrated BCE and BP dates, indicating the main archaeological periods and/or cultural entities.

(Rossignol-Strick 1995, 1997). This chronological discrepancy seems mainly to arise from the effects of hard water, the influx of old carbonates into lakes and their incorporation within the same organic-rich sediments that are commonly used for dating (e.g. Baruch 1994; Baruch & Bottema 1999; Cappers *et al.* 1998; van Zeist & Bottema 1991).

    The general course of the palaeoclimatic changes in the Near East, often determined as a sequence of fluctuations between wet and dry phases, corresponds to the palaeoclimatic sequence of the North Atlantic, in spite of some discrepancies between the available radiocarbon dates from the various sources (e.g. Bar-Mathews *et al.* 1997; Bar-Mathews *et al.* 1999; Frumkin *et al.* 1999; Henry 1997; Sanlaville 1996, 1997). Each of the climate-sensitive sources indicates abrupt climatic fluctuations within the long trend of change. The entire sequence can be summarized as follows.

During the Last Glacial Maximum (LGM, *c.* 24,000 to *c.* 18,000 BP) the climate of the region was cold and dry, although the coastal hilly and mountainous ridges enjoyed winter rains and were covered mostly by open forests. The parkland belt of terebinth and almond stretched over the eastern and northern margins, bordered by steppic and desertic vegetation. Numerous inland lakes shrank and some even turned into salty ponds. From 18,000 BP onwards, precipitation across the region slowly increased, and even more rapidly from around 15,000/14,500 BP, marking the onset of the Bölling–Allerød warmer period. The latter ended *c.* 13,000/12,800 BP with the onset of the Younger Dryas (YD). Coeval was the increase in the amount of CO_2 in the atmosphere, which generally favours annual plants with C3 pathways, such as the cereals (Sage 1995). A drop in the average annual temperature, a decrease in the availability of atmospheric CO_2, and the decrease of rains marked the YD. The reconstructed vegetational map for this period by Hillman (1996) shows the impacts of the new conditions. These are marked by the retraction of the vegetational belts and, although the decrease of CO_2 is less effective when trees are considered, the decrease in precipitation was probably more important. The end of the YD came with the climatic improvement at the onset of the Holocene (some 11,500 BP), which marked the expansion of the woodlands and steppic belts inland.

The return to pluvial conditions began around 11,500/11,300 BP. Rainfall in the southern Levant may have reached the levels of the previous peak (Bar-Mathews *et al.* 1997) and it also increased in Anatolia, penetrating gradually into the Zagros Mountains during the Early Holocene (Hillman 1996; van Zeist & Bottema 1991). The Early Holocene climate was moister than that of today, as is indicated by climatic proxies from the northern hemisphere such as the ice cores and various pollen records. An additional major climatic crisis is currently dated to 8,400–8,200 BP (Alley *et al.* 1997; Bar-Mathews *et al.* 1999). Not surprisingly, these centuries mark the end of the Pre-Pottery Neolithic B (PPNB).

The effects of the monsoon system were limited to a region encompassing a large part of the Arabian peninsula and portions of north-east Africa, as recorded in the numerous palaeo-lake deposits and Late PPNB occurrences (Hassan 1997). This meant that for a couple of millennia, certain areas of the Sahara and Arabian Desert enjoyed both winter and summer rains.

Sea-level rise after the LGM was gradual and continued until the Mid-Holocene, as evidenced by submerged PPNB and Early Pottery Neolithic sites uncovered along the Israeli shoreline (e.g. Galili & Nir 1993). Independent of regional tectonic conditions, evidence for the same sea rise was recently obtained in the western Mediterranean (Lambeck & Bard 2000). During the Holocene, a stretch of land around 2 to 40 km wide and 600 km long from south-east Turkey to northern Sinai was lost. The coastal inundation affected the size of territories of both foragers and early farmers. Their vulnerability as a result of their dependence on the natural food sources meant that hunter-gatherers were definitely affected much more than later farmers. The shifting by the Natufians of the location of the main base camps from the central area of the coastal plain to the foothills, as will be described below, marks the shift of the ecotone. In this regard, it is worth mentioning that aquatic sources were always of minimal economic importance because the eastern Mediterranean Sea is and was more saline in comparison with its western side or the vicinity of the Nile Delta. Hence, the main impact of the sea-level rise could have been on the availability of particular marine shells such as *Dentalia*, which were used for decoration and often collected on the beaches.

In sum, in a geographically varied region such as the Near East, palaeo-climatic fluctuations could have played a major role in the spatial distribution of resources. Hence, evolutionary models that predict the social evolution in this region from mobile or semi-sedentary foragers to the emergence of seden-tism must take into account the shifts in the reliability, predictability, and accessibility of resources.

LATE PLEISTOCENE FORAGERS AND THE EMERGENCE OF INEQUALITY

The archaeology of Late Palaeolithic (or Epi-Palaeolithic, *c*. 20,000–15/14,500 BP) foragers is relatively well known (e.g. Bar-Yosef & Belfer-Cohen 1989a, 1992; Bar-Yosef & Meadow 1995; Byrd 1994a, 1998; Fellner 1995; Garrard 1998; Garrard *et al.* 1994; Goring-Morris 1987, 1995; Goring-Morris & Belfer-Cohen 1997, and references therein; Henry 1989, 1995; Rabinovich 1998). Social entities within this world of foragers were identified on the basis of quantitative and qualitative stylistic differences among the microlithic tool types, which changed faster than the core reduction strategies (Goring-Morris *et al.* 1998). Cultural attributes such as spatial distribution of sites, site size and structure, intensity of occupation, and patterns of seasonal mobility are the components that facilitate the configuration of a socioeconomic map. In a west–east transect in the Levant, or south–north in Anatolia, semi-sedentary foragers occupied the Mediterranean vegetational belt while more mobile groups subsisted in the parkland, the steppic areas bordering the Syro-Arabian desert, or the high Anatolian plateau.

The main cultural change is marked by the appearance of what is called the Natufian culture (Figure 3), which is traditionally divided on the basis of stratigraphies and radiocarbon measurements into Early and Late Natufian (*c.* 15,000/14,500–11,500 BP: e.g. Bar-Yosef 1998; Belfer-Cohen 1991b; Valla 1995). The Early Natufian is known for the various aspects of hamlet or village

Figure 3. The Natufian homeland with additional, contemporary sites.

life, an interpretation based on the uncovering of pit-houses, burials, imagery objects, and numerous pounding and grinding stones, as well as rich lithic and bone industries. The full array of the cultural components characterizes the larger Natufian sites (*c.* >1000 m^2) in the homeland in the central Levant (Figure 3), and rarely the small (15–100 m^2) seasonal camps (Bar-Yosef 1998, and references therein; Bar-Yosef & Belfer-Cohen 1992; Belfer-Cohen & Bar-Yosef 2000; Byrd 1994a; Henry 1989).

It is the presence of commensals, as mentioned above, that defines the Early Natufian hamlets as sedentary (e.g. Tchernov 1991b, 1993a). The cause of Natufian sedentism is still debated. Henry (1989) advocated that it was brought about by intensification of cereal exploitation. Alternatives include the environmental impact caused by a short and cold period (Dryas II) on the increasing population of foragers. There was also the climatic amelioration of the Bölling–Allerød, and the attraction of stable and expanding vegetal resources (e.g. Bar-Yosef & Belfer-Cohen 1991; Garrard 1999). In addition, McCorriston and Hole (1991) proposed that sedentism enhanced the propagation of annuals such as cereals. However, Early Natufian sedentism lasted only for about two and a half millennia, and site stratigraphies testify to intervals of abandonment, the lengths of which are unknown (Valla 1991; Valla *et al.* 1998). These short gaps may constitute the evidence for the cyclical nature of sedentism.

The Natufian economy is better known from animal bones than from plant remains. The latter are rare in *terra rossa* soils, but are present in loamy and ashy deposits such as in the lower layers of Abu Hureyra and at Mureybet (Colledge 1998; Hillman *et al.* 1989; van Zeist & Bakker-Herres 1986). Wild cereals, legumes, and fruits such as almonds and pistachios were gathered with other numerous species. The older, but well-preserved assemblage of Ohalo II, dated to *c.* 21,000 BP, indicates that the full vegetarian menu of the Natufian (Kislev *et al.* 1992) was available several millennia earlier. However, the main difference was that the Natufians harvested the cereals with sickles. This new tool type was composed of shaped flint blades, which were inserted in grooved bone or wooden handles. Experimental and microscopic studies (Anderson 1998; Unger-Hamilton 1991; Yamada 2000) demonstrate that the sickles were mainly employed in cutting cereals. Their use, instead of that of basket and beaters, can be interpreted as showing either that higher yields per harvested area were thereby obtained or that the gathering was done when most ears were still green and the seeds were thus more firmly attached. Harvesting at this earlier stage required parching as a major step in food preparation. The scarcity or lack of storage facilities in Natufian sites is intriguing. Ain Mallaha is the only site where underground pits, partially coated with plaster, were reported (Perrot 1966). Instead, it seems that baskets fulfilled this need. The presence of baskets is indicated by use wear on bone tools (Campana 1989). Baskets probably served for transporting seeds and fruits to the hamlet, as well as for storage. Basket remains from PPNA and PPNB contexts hint that it is only a matter of time until similar finds are discovered in a Natufian site. Finally, the overall exploitation of plants and animals reflects a 'broad spectrum' subsistence. Hunting and trapping of mammals, such as rare aurochs, numerous gazelles, fallow deer, roe deer, and hares, and of reptiles such as tortoises took place, as did some fishing (e.g. Davis 1982; Tchernov 1993b). Hunting tools probably included bows and arrows. Grooved 'shaft straighteners' made of basalt and limestone bear burn marks that resulted from rubbing heated wooden shafts and these support the notion that the Natufians already knew archery (e.g. Valla 1987).

Kitchen equipment included bedrock mortars, portable mortars, bowls of various types, cup-holes, mullers, and pestles. Many small mortars and numerous pestles were made from basalt. The latter originated from the lava outcrops in the eastern Galilee and the Golan, some 60–100 km away from Mt Carmel (Weinstein-Evron 1997; Weinstein-Evron *et al.* 1995). Unique boulder mortars weighing up to 100 kg and with a 70–80 cm deep hole — sometimes called 'stone pipes' — were considered as special tools and have often been found in an upright position, with breached base, only in graves. One possible explanation was their use as a means for communicating with the dead.

The Natufians used a large number of bones and horn cores to shape various objects. Many pieces were employed in daily tasks such as hide-working and basketry (Campana 1989). Barbed items served as parts of hunting devices (spears or arrows). Hooks are few, but gorgets are numerous, and both are assumed to have been made for fishing.

Large collections of unique bone implements decorated with incised patterns or carved animals (Bar-Yosef & Belfer-Cohen 1998) are often found in special contexts, such as in Hayonim and El-Wad caves. The special stone structures in both caves (although they are not well recognized at El-Wad) are interpreted as locations for particular activities, perhaps carried out by the shamans. Hence, unique objects, including the decorated sickle hafts with three-dimensional carved young ungulates, were found there. Other special finds are the limestone slabs uncovered in Hayonim cave (Belfer-Cohen 1991a). These were often incised with the 'ladder' pattern interpreted as notation marks (Marshack 1997). A recently found small slab depicts distinct incised units made by a series of parallel lines. Their overall arrangement conveys the impression of territories or 'fields' of some kind (Bar-Yosef & Belfer-Cohen 1999). This interpretation corroborates the material evidence for 'localization' of Natufian groups mentioned above, which probably signals their perception of 'territorialism' and ownership. Different incised slabs with a carved meander pattern were uncovered in probably a domestic context, in Wadi Hammeh 27 (Jordan). The wavy pattern, which may symbolize water, is a motif also known from large carved basalt bowls from Mallaha and Shukbah.

The contents of the graves, as mentioned above, are taken as indicators for social inequality during the Early Natufian. In most cases, graves were dug in deserted pit-huts, in open spaces, and not under the floors (Belfer-Cohen 1995; Byrd & Monahan 1995). The relationship between the living and the dead is therefore less straightforward than was previously considered. Grave pits were shallow or deep, and rarely were paved with stones or lime-coated. The burials, which reflect variable mortuary practices, include supine, semi-flexed or flexed positions with various orientations of the head. The number of inhumations varies from single to collective, although the latter are more common in Early Natufian contexts. Secondary burials were either in special graves or mixed with primary burials. About 8 to 10 per cent of Early Natufian skeletons had body decorations of marine shells (mainly *Dentalia*), and bone and animal teeth pendants which mark the remains of garments, belts, headgear, and the like (Belfer-Cohen 1995; Byrd & Monahan 1995). In a few cases, objects recognized as grave offerings were found. In addition, joint inhumations of humans and domestic dogs occurred in two graves, one at Ain Mallaha and one in Hayonim Terrace (Davis & Valla 1978; Tchernov & Valla 1997; Valla 1990).

The decorated burials raise the issue of social status within Early Natufian communities (Wright 1978). Although recent analyses do not support the contention that they reflect a ranked society (Belfer-Cohen 1995; Byrd & Monahan 1995), the fact that in each site the decorated skeletons of individuals of all ages form only a portion of the entire buried population suggests otherwise. In addition, the differences in the composition and types of the common elements of body decoration (bone beads and pendants) between the sites is interpreted as marking the identity of particular local groups (Belfer-Cohen 1995).

A cultural change is observed in Late Natufian burials, such as in the large cemetery at Nahal Oren (Mt Carmel), where decorated burials are lacking and only a few grave goods were found (Belfer-Cohen 1995; Byrd & Monahan 1995). Moreover, there are more secondary burials in these contexts, as well as a few cases of skull removals, which herald a custom common in the ensuing Early Neolithic period (Belfer-Cohen 1988).

Most authorities agree that Late Natufian society was more mobile in comparison with its ancestors, probably because of the harsher and less stable environmental conditions created by the Younger Dryas. Under these conditions, mortuary practices convey a shift from a non-egalitarian society back to a more egalitarian one (Kuijt 1996).

During both social phases of Natufian society, long-distance connections or even alliances are indicated by the presence of marine shells. Most items were collected from the shores of the Mediterranean Sea, and some were brought from the Red Sea (D.E. Bar-Yosef 1989, 1991; Reese 1991). Among the rare, but most informative finds are a freshwater clam, *Aspatharia* sp., which originated in the Nile Valley, and a few pieces of Anatolian obsidian from the upper layer at Ain Mallaha.

Among the mobiliary art objects, the shift from the non-egalitarian Early Natufian to a simpler society in the Late Natufian leads to the increasing presence of personal items (known mainly from the burials at Nahal Oren). Among these, it is worth noting the double-headed objects with carvings at the extreme ends of each item depicting an owl and a dog's head, or an ungulate and a human head. A rare item is the 'baboon' head, which may hint at an African connection. All portable objects could have served as transferable items, thus acquiring an additional symbolic value.

THE YOUNGER DRYAS AND THE INITIATION OF INTENTIONAL CULTIVATION

The climatic crisis of the Younger Dryas, mentioned above, is best known from the climatic proxies gathered in the northern hemisphere (e.g. Broecker 1999; Zolitschka *et al.* 1992). The length of this period, according to the ice core chronology, is *c.* 1,300±70 years (Alley *et al.* 1993; Mayewski & Bender 1995; Taylor *et al.* 1997), from 12,900 to 11,600 or 12,800 to 11,500 BP. However, there is a certain discrepancy between the European varve chronology, the calibrated radiocarbon and ice core dates and local records such as in Lake Van, where varve counting suggested dates of 12,600 to 11,000 or 10,510 BP (Lemcke & Sturm 1997). In spite of these reservations, what concerns us is the reaction of human societies to an abrupt climatic change, even if, when filtered through the environment, the effects seem to be slow on the scale of a human lifetime. The case of the Natufian is probably the oldest for which we can test the relationship between environmental and cultural changes in the Near East.

Complex hunter-gatherer societies such as the Early Natufian lasted for two and a half millennia, but are considered as unstable social entities (Arnold 1996, and references therein; Henry 1991). It seems that the crisis of the YD, by imposing drier and colder conditions, caused populations of many but not all hamlets to disperse and become more mobile. Apparently, a sedentary mode of production became incompatible with the previous trend of population growth, especially among those who occupied the marginal belts of terebinth/almond woodland and steppic vegetation. Under conditions of intense competition, cost considerations favour smaller groups, which are economically advantageous when marginal returns diminish (Belfer-Cohen & Bar-Yosef 2000; Kosse 1994). Therefore, the Natufian and other social entities in the Near East faced a variety of choices (Bar-Yosef & Belfer-Cohen 1991), and the archaeological records reflect the implementation of their decisions. In the Natufian homeland, increasing mobility between base camps was a partial solution. The greater mobility of the Late Natufian marks a return to egalitarian society and is indicated by the disappearance of decorated burials and the larger number of multi-individual graves. Kuijt (1996) considers this shift in mortuary practices an attempt to erase social differentiation between kin-groups and to emphasize the unity of the population by giving similar treatment to all community members. In addition, skeletal data indicate conditions of physical stress (Belfer-Cohen *et al.* 1991).

Natufian groups in the steppic areas of the Negev and northern Sinai developed a special adaptation labelled archaeologically as the Harifian culture (Bar-Yosef 1987; Goring-Morris 1991). The exploitation strategy of the Harifian incorporated extensive seasonal movements between lowlands in the winter and highlands in the summer and autumn. These groups also enjoyed an enhanced exchange relationship with their neighbours, reflected in the marine shell jewellery, which was collected from both the Mediterranean and the Red Sea shores. In spite of their efforts, this cultural entity lasted only a few centuries. Despite ambiguities in interpreting the calibrated dates, it is probable that the last Harifians died or joined the PPNA cultivating communities in the Jordan Valley.

A different socioeconomic solution emerged in the face of the worsening conditions in the Taurus and Zagros foothills. On the Tigris tributary, the hamlets of Hallan Çemi Tepesi (12,900–10,900 BP) and further east Zawi Chemi Shanidar in Iraqi Kurdistan indicate that sometime between *c.* 13,000 and 11,200 BP groups shifted from mobile hunting and gathering to semi-sedentary settlements. Their rounded pit-houses do not seem to differ from those of the Natufian (Rosenberg 1998; Solecki 1981), but their lithic industry, rich in triangles, ties them with the Trialetian culture, further north (Kozlowski 1999). The maintenance of long-distance connections is indicated by the chlorite stone bowls, which were produced and transported along the eastern Taurus and Zagros foothills (Aurenche & Kozlowski 1999). In addition, the resemblance of the architectural remains of the pit-houses of Hallan Çemi to those of the Early Natufian hamlets, as well as the presence of an open public space, indicates that it was a village of a non-egalitarian social group.

THE FIRST FARMING VILLAGES

The current archaeobotanical remains bear witness, during the later centuries of the Younger Dryas or immediately after, to the onset of intentional cultivation of wild cereals and legumes. It may be hypothesized that it was due to the annually reduced yields in the natural stands of einkorn and emmer wheats, rye, and barley that those groups which had previously exploited these 'r-resources' and knew the nature of the species began planting them. The reduction in wild cereal production could have been the result of decreasing atmospheric CO_2 or a response to increasing human demands. Such a stressful situation was created when natural fields shrank under the cold and dry conditions of the Younger Dryas.

Colledge (1998) analysed the relationship between the occurrence of the various plant taxa and their ecological classification by applying correspondence analysis to available carbonized plant assemblages. That from PPNA Mureybet led her to conclude that cultivation of wild cereals was practised near the site. Similar conclusions concerning wild barley were reached by Kislev (1997) in reporting the plant collection from Netiv Hagdud, a PPNA site in the Jordan Valley. Hence, because of a lack of sufficient evidence, it is as yet uncertain whether the initiation of cultivation took place within what is archaeologically defined as the Late Natufian and its contemporaries (such as in Abu Hureyra I), or in the Khiamian.

The Khiamian is the first Neolithic archaeological manifestation in the Levant (Figure 2). This entity is still poorly known, perhaps because of its short timespan of barely a few centuries (c. 11,700–11,200 BP: Aurenche & Kozlowski 1999; Kozlowski 1999; Kozlowski & Gebel 1996). In addition, the available information on the Khiamian was obtained from very limited soundings and sites in which a mixture with earlier layers is likely to have occurred. The lithic industry of the Khiamian comprises of the aerodynamically shaped el-Khiam arrowheads, asphalt-hafted sickle blades, some microliths, and high frequencies of perforators. Bifacial or polished celts, considered Neolithic 'markers', are absent from the reported contexts.

Early Neolithic settlements (Figure 4) are better known from the Jordan Valley, the Damascus Basin, the Euphrates Valley and beyond (such as Qermez Dereh: Watkins *et al.* 1989), or the lower level at Çayönü (Özdogan 1999). Most, but not all of these hamlets and villages (0.2 to 2.5 hectares in size) are three to eight times larger than the largest Natufian sites, thus reflecting a rapid population growth.

Three cultural entities of farmers-hunters were identified in the Levant, including the Mureybetian in the north, Aswadian in the centre, and the Sultanian in the south. In addition, foragers continued to survive in Anatolia as well as in the semi-arid areas of the Syro-Arabian desert, the Negev and Sinai. The domestic architecture of farming communities is characterized by pit-houses with stone foundations and walls built of plano-convex unbaked mud-bricks or adobe. In time they became rectangular (Stordeur *et al.* 1997). Storage facilities are found in every site, either as small, stone-built bins or larger, mud-brick constructed installations (Bar-Yosef & Gopher 1997; Cauvin 1977).

In the Neolithic contexts, special buildings, carved stelae, mortuary practices, modelled statues, and figurines are considered as indicators of ceremonial

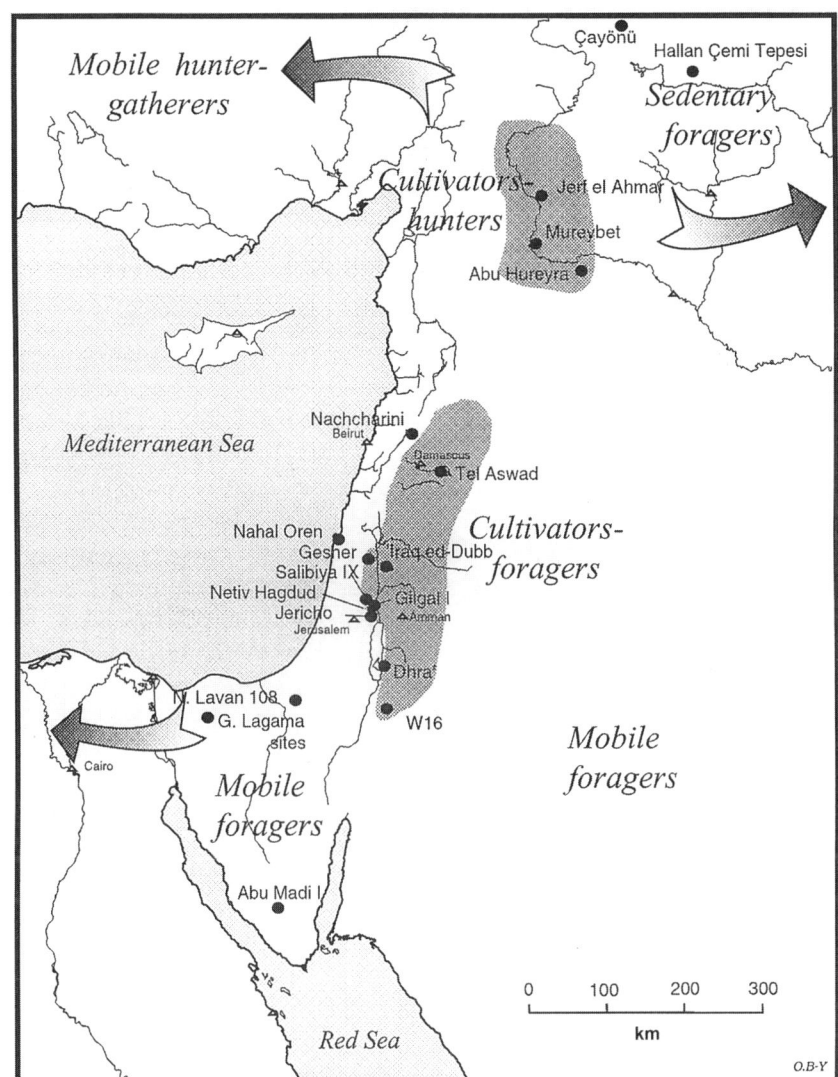

Figure 4. The Early Neolithic of the Levant. The stippled areas show the core regions of early cultivation. The arrows mark the geographic advent of the Neolithic Revolution.

centres, belief systems, and rituals. One of the first signs of emerging social complexity is communal building efforts, such as the walls and the tower of Jericho, orchestrated by a leader or headman. These architectural elements were interpreted by Kenyon (1957) as parts of a defensive system against raids by human groups. Unfortunately, Kenyon disregarded the fact that a tower, as an integral part of a defence system, must be built on the outer face of the walls to enable the defenders to shoot sideways at the climbing attackers. An alternative interpretation (Bar-Yosef 1986) suggests that the wall was erected on the western side of the site to protect the settlement against mudflows and flash floods. In addition, there is a sound basis to the proposal that there was probably only one tower in Jericho. Although its function is unknown, it probably accommodated a small, mud-brick shrine on top. True, unequivocal evidence for public ritual is missing, but the open space north of the tower (Area M) may have been similar to the 'plaza' in Çayönü (Turkey), which served for public gatherings (Özdogan 1999).

In the Sultanian of the Jordan Valley, most burials are single, with no grave goods. Skull removal was performed usually only on adults, while child burials were left intact. Isolated crania are sometimes found in domestic areas or special-purpose buildings. The differentiated treatment along age lines reflects changes in attitudes towards the dead within Early Neolithic society (Kenyon & Holland 1981; Kuijt 1996). It seems that greater social value was attributed to adults, as evidenced by the conservation of their skulls, while children were not valued in the same manner. This is rather surprising in a society where additional workers were needed. It also marks a departure from the Natufian tradition.

The appearance of human figurines shaped from either limestone or clay along gender lines is a cultural novelty. They depict either standing or kneeling females. Those classified as the 'seated woman' type may herald more elaborate manifestations of the same subject in the succeeding PPNB civilization. Common interpretations view the explicit expression of gender, which was not evident in the Natufian, as indicating the emerging distinct role of women in a society of farmers-hunters. Some suggest that this shift brought about the cult of the 'mother goddess' in later centuries (Cauvin 1985).

Finally, an important shift from the Natufian tradition of food preparation is gleaned from the abundant pounding tools, including flat slabs with cupholes, rounded shallow grinding bowls, and hand stones, often loaf-shaped. Only rare mortars or deep bowls reflect the previous tradition of heavy-duty kitchen equipment. The full array of consumed food is known from the high frequencies of carbonized seeds of barley, wheat, and legumes. Current research favours the interpretation that the seeds were obtained mostly through intentional cultivation. The debate concerning the morphological features which traditionally were considered as attributes of domesticated species seems to favour the interpretation that the harvested species were still mostly wild (Colledge 1998; Kislev 1989, 1997; Zohary 1992). Moreover, Early Neolithic villages departed from Natufian locations in numerous cases and preferred the alluvial soils of fans and river terraces.

Worth mentioning is the fact that Early Neolithic villagers continued to gather wild fruits and seeds, and to hunt and trap. Gazelle, equids, and cattle were hunted in the middle Euphrates area, while gazelle, fox, a few fallow deer, wild boar, and wild cattle were the main game animals in the Jordan Valley (e.g. Tchernov 1994). Large numbers of birds, especially ducks, were trapped by the occupants of all sites. Lizards and tortoises were also gathered. The overall picture is one of a 'broad spectrum' subsistence strategy, similar to that of the Natufians.

The desire for foreign commodities is witnessed in the long-distance importing of obsidian found in several villages in the southern Levant, some 500 km south of its source area in central Anatolia. However, not all settlements were able to obtain such precious goods. Marine shells were brought from the Mediterranean coast, with fewer coming from the Red Sea. There is a clear shift in the type of shells selected for exchange. *Glycymeris* and cowries become important, but *Dentalia* shells (yielded where excavated deposits were sieved) were still in use (D.E. Bar-Yosef 1991).

In sum, the archaeology of PPNA sites clearly demonstrates the emergence of a non-egalitarian agricultural society, which continued to rely on hunting and gathering. Signs of social ranking are expressed in mortuary practices, frequencies of foreign imports such as obsidian, and the appearance of the first public structures that testify to communal ceremonies.

THE PPNB CIVILIZATION

Excavations of PPNB sites (defined on the basis of radiocarbon dates and lithic assemblages) in the Levant and Anatolia during the past decade have uncovered the presence of a major civilization. Large village sites, elaborate building techniques, ceremonial centres, carved stone stelae, and the like demonstrate that the 'PPNB civilization' deserves to have a different status in our archaeological literature and social interpretations.

In each large village, domestic buildings reflect the basic social units. Thus nuclear families probably occupied the rectangular houses that were later subdivided into smaller rooms, while extended families shared accommodations in compounds such as those in Bouqras (Akkermans *et al.* 1983). Houses with two storeys were more common in the later part of the PPNB, reflecting both population growth and increased material richness. Among these are the corridor houses in Beidha and the well-preserved two-storey houses of the 'cell' type in Basta. The existence of larger houses alongside smaller ones expresses unequal wealth and social status. These are only known from sites where the excavated area is large, such as in Çayönü, where over 5,000 m^2 were exposed (Özdogan 1999; Özdogan & Özdogan 1998).

The presence of ceremonial areas or special buildings for rituals that served as shrines is now recognized in several sites, and in particular if the edge of the village was uncovered. Examples include Çayönü, Nevali Çori, and Beidha (Byrd 1994b; Kirkbride 1968; Özdogan & Özdogan 1998). In addition, within each of the 'tribal territories' we may expect a sacred settlement, such as Göbekli Tepe (Hauptmann 1999; Schmidt 1999) and Kefar HaHoresh (Goring-Morris *et al.* 1995). It is not impossible that the site of Çatalhöyük, or a certain portion of it, served the same purpose (e.g. Hodder 1999). In each of these sites the archaeological context reflects both domestic and ritual activities, forming a continuum from the sacred to the mundane. The sculptures at Göbekli Tepe and Nevali Çori, as well as the reconstructed buildings at Çatalhöyük, exemplify in their animal and human depictions the complexity of symbols, which are not easy to decipher.

No less important are the caches of plaster statues depicting human figures uncovered in Jericho and 'Ain Ghazal (Rollefson 1983). Their archaeological context testifies to the intentional burial of used cultic objects (Garfinkel 1994). The breakage of such holy items prior to their interment is a well-known phenomenon from historical periods in the Near East. The interpretation of the plaster statues, some of which are only busts, is not easy. According to the position of their hands, those holding the lower part of the belly are considered female representations (Aurenche & Kozlowski 1999). All have eyes encircled with asphalt lines, and stripes of red colour on their bodies. By employing archaeological analogy to later millennia, the statues seem to represent a pantheon of deities, in which the human figure represents both the real and the mythological image. Amiran (1962) suggested that the mode of production of these statues — which were constructed with reeds, cloth, and plaster — resembled the creation of Man as depicted in the Gilgamesh epic. Hence, it is quite probable that the cosmology of the PPNB civilization, orally transmitted in the Near Eastern world, found its written expression several millennia later.

The territories of the kinship-based entities (tribes?) were marked by sacred localities that symbolize, in a similar way to Sheikh tombs in southern Sinai (Marx 1977), ownership of the land. Such a special locality is the dark cave filled with paraphernalia in Nahal Hemar. The location of the site marks the geographic boundary between the Judean hills and desert and the northern Negev. It seems that each area was occupied by a different Neolithic group. Among the objects in the cave are skulls modelled with asphalt mixed with collagen, stone masks, and small human figurines, as well as other broken items (Bar-Yosef & Alon 1988; Bar-Yosef & Belfer-Cohen 1989b).

The Levantine PPNB knapping operational sequences are employed, together with the radiocarbon chronology, in order to subdivide the period into phases and to distinguish territorially defined groups (e.g. Bar-Yosef 1981; Cauvin 1978, 1997; Cauvin & Stordeur 1978; Gopher 1994a; Quintero & Wilke 1995; Wilke & Quintero 1994). Because of the large size of the common arrowheads, they are considered as markers of what S. Kozlowski named the Big

Arrowhead Industry (Aurenche & Kozlowski 1999; Kozlowski 1999). Their geographic expansion marked the advent of PPNB communities into Anatolia, thus establishing a different historical trajectory from the one that took place in the Zagros foothills (Figure 5). In the latter region (mostly in western Iran) the production of microliths by the local farmers continued through Early Neolithic times (Hole 1994; Kozlowski 1994). Hence, we can suggest schematically that, while the Neolithic villages in the Anatolian plateau were established by colonizers, the inhabitants of the Zagros foothills adopted the new food-producing techniques and a suite of crops.

Assuming that the chronological determinations are correct, the PPNB civilization emerged in the northern Levant, which is geographically a larger area than the southern part. Thus, the cultural variability that is expected between the eastern Taurus foothills, the Amuq basin, the Euphrates River and its tributaries (the Balikh and the Khabur), and the upper Tigris Valley is greater than in the southern Levant. In addition, this province was also the area in which interactions between the foragers beyond the Taurus and the Zagros foothills occurred and therefore certain sites preserve the mixed characteristics

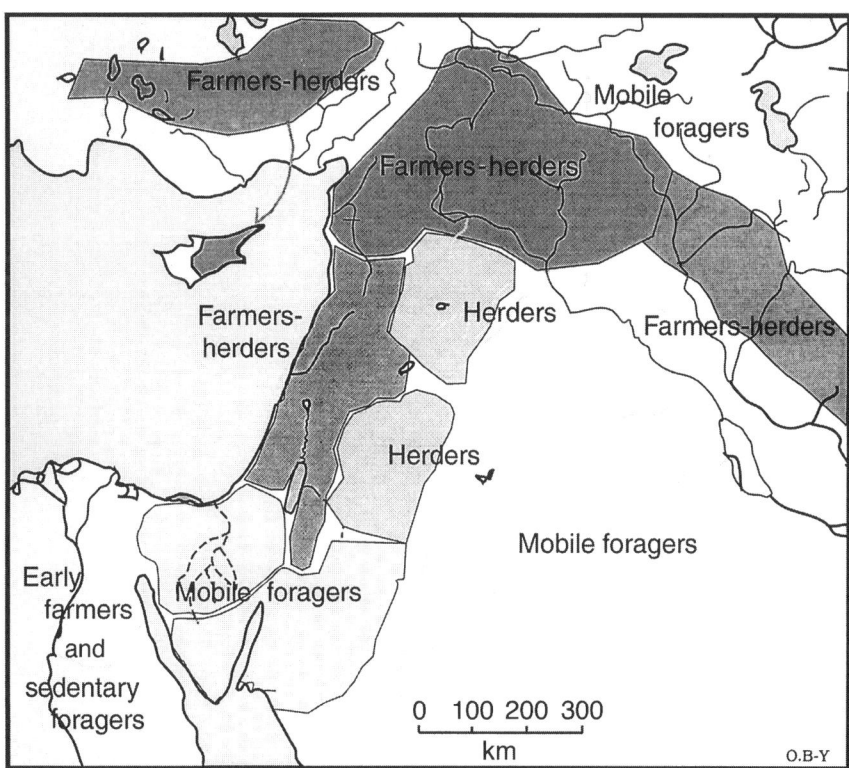

Figure 5. The territories of the various entities within the PPNB civilization.

of both cultural spheres (e.g. Bar-Yosef 1997a; Cauvin *et al.* 1999; Kozlowski & Gebel 1996).

The economy of the PPNB settlements was based on the full suite of annual crops such as barley, wheat, rye, flax, and legumes (later including broad beans and chick peas), which were already domesticated (e.g. Aurenche & Kozlowski 1999; Garrard 1999). Storage facilities were common (Kuijt 2000b). Animal bones reflect the domestication of goats and sheep, which were the predominant game of foragers in the Taurus–Zagros ranges. At least in the first millennium of the PPNB, the tended animals do not demonstrate the change in size that was previously seen as a morphological marker for domestication (Hesse 1984; Legge 1996; Vigne *et al.* 1999; Zeder & Hesse 2000). Following the primary penning and herding of goats and sheep, both were introduced into villages of the central and southern Levant (Garrard *et al.* 1996). It was only during the later times of the Pottery Neolithic period that herding was adopted by the inhabitants of the steppic belt, and not before the early Chalcolithic that fully fledged pastoral nomadism became a common way of life in the Levantine deserts.

The domestication of pigs and cattle followed that of goats and sheep. The penning of pigs seems to have been tried at earlier times in Hallan Çemi (Rosenberg *et al.* 1998). It is not before the Middle and Late PPNB that the faunal evidence testifies to the intentional raising of pigs, mostly in the more humid areas of the north (such as Çayönü) and the coastal Levant (Bar-Yosef & Meadow 1995).

Cattle domestication seems to have taken a different course. There is general agreement that, at least in the case of south-west Asia, the incorporation of the aurochs was motivated by religious reasons no less than simply by basic dietary needs. This interpretation is based on the contexts in which cattle remains were uncovered in various sites in the northern Levant and Anatolia. Skulls with horns were found intact in dwellings and trash pits, and are known from the by now famous examples of the Çatalhöyük buildings (e.g. Hodder 1999; Mellaart 1967).

PPNB farming communities were flourishing and expanding. In the Levantine region this is seen in the size of sites, which range from 2.5 to 12 hectares. Similar sizes were recorded in the Anatolian province in Asikli Hüyük and Çatalhöyük. Among the sites themselves there is a clear size hierarchy. Although ethno-archaeological studies (e.g. Kramer 1983) demonstrated that surface area cannot be translated by a simple formula into the number of inhabitants, we consider the measures of hectares as providing a relative scale that exemplifies differences in population sizes. Assuming that the largest tested sites accommodated a viable biological unit of about 400–500 people, it could mean that 'tribal' territories (Figure 5) were inhabited by about 1,500–2,500 people.

In sum, when the evidence for site size, public ceremonial and domestic ritual activities, mortuary practices, and amount of transportable commodities, as well as other features, are taken into account, the comprehensive list demonstrates the variable degrees of social complexity within the PPNB civilization and its various tribal territories. While within certain regions human societies could have enjoyed a more egalitarian structure, the evolving inequality within most areas made these Neolithic populations ready for the emergence of chiefdom. The climatic disaster *c.* 8,400/8,200 BP resulted in a major collapse.

THE PPNB FORAGERS

The importance of the PPNB civilization is particularly noticeable in comparison with what happened in the geographically marginal areas which had been an integral part of the larger interaction sphere (Bar-Yosef & Belfer-Cohen 1989a; Cauvin 1997). At the edges of the PPNB *koine,* Neolithic sites contain rounded pit-houses, rich assemblages of lithics with numerous arrowheads, and grinding stones. The fauna reflects hunting in the local environment and the plant remains, where preserved, testify to gathering. These were the remains of the contemporary foragers (Bar-Yosef 1984; Garrard *et al.* 1988, 1994). The occupied sites are of various sizes, up to 1,000 m^2. Certain sites, such as Ujrat el Mehed in southern Sinai, revealed evidence for ritual secondary burials of adults (complete with skulls) in underground storage facilities (Hershkovitz *et al.* 1994), thus representing a belief system different from that of the farmers. This site occupies a topographic location that could have been a focal point for annual aggregations, a proposal supported by the lithic analysis, which demonstrated that most if not all the lithics were brought in. About 20,000 pieces, including 6,000 projectile points, were found with only two dozen cores (Gopher 1994b). Other curated objects include large collections of marine shells, which were mostly shaped to serve as both items of jewellery and elements of barter and exchange (Bar-Yosef Mayer 1997).

Human societal interactions are constantly changing. These prehistoric societies of farmers and foragers, especially as they differed in their economic base, could have maintained amicable relationships, which may have led to intermarriage, but could infrequently have led to physical conflict. An interesting aspect, as yet not fully explored, is the role of mobile foragers in implementing trade or exchange. The presence of marine shells from the Red Sea among the inland farming communities and the 'down the line' movement of obsidian from central Anatolia into the Levant, to mention but two examples, could have been in part accomplished by the more mobile groups of hunter-gatherers.

In this respect the special type of game drives known as the 'desert kites' are an important feature. These were probably laid out by PPNB foragers in order to hunt *en masse* (Meshel 1974). Employing such a technique, by which numerous gazelles and onagers were hunted, could have been the response to demands for extra traded-in meat by large farming communities, who capitalized on their flocks. In one case, the foundation of a rectangular house in a foragers' camp (Garrard *et al.* 1988) littered with rounded pit-houses could be interpreted as the 'merchant's temporary home'.

From peaceful interactions, we move to the possibility of intercommunal conflicts. As argued by L. Keeley (1996: 18), archaeologists during the past three decades 'have increasingly pacified the past', mostly because concrete evidence for warfare has not been recovered. Even the suggestion mentioned above, that the PPNA Neolithic tower in Jericho and the wall on its outside perimeter were not elements of a fortification scheme, was taken to mean something that was not originally intended (Otterbein 1997). Among PPNB sites only one example of the burning of an entire village was recorded (Ganj Dareh: Smith 1976), for which the reason is unknown. Pre-state societies engaged from time to time in warfare for various motives such as for obtaining booty, vengeance, and glory, but not for political control. This type of warfare could sometimes have characterized farmer–forager interactions in the context of emerging social complexity and concentration of wealth among the PPNB tribal societies as demonstrated above.

THE COLLAPSE OF THE PPNB CIVILIZATION

Stratigraphical unconformities and temporary site abandonment were not uncommon during the PPNB period. Only very few settlements survived for many centuries. Various reasons account for the abandoning of houses in a living village, from the death of the head of the family to the outcome of verbal and physical conflicts (e.g. Cameron & Tomka 1993, and papers therein). However, when the entire village is deserted the reasons could be more complex, from over-exploitation of the immediate environment or societal conflicts to the impact of consecutive droughts. Under any circumstance, the abandonment of one site and/or several may precipitate societal restructuring, especially among farming communities such as those of the PPNB. It is therefore imperative first to document the timing of abandonment, and whether it was only local or encompassed an entire region. We then need to search for the reasons, which, as with every inquiry into the 'why' question, are open to disagreements.

The stratigraphic gap between the PPNB layers and those labelled as Pottery Neolithic is well established in the Levant and eastern Anatolia (Aurenche & Kozlowski 1999; Gopher & Gophna 1993). Further support for the observation concerning the abandonment of PPNB villages was gleaned from the establishment of new hamlets and farmsteads in the southern Levant which contained characteristic ceramics and a lithic industry identified as the Yarmukian (Garfinkel 1993, 1999; Stekelis 1972).

Worth noting is that the presence of pottery production, which began during the Late PPNB in the northern Levant, reached the southern Levant only in the so-called 'Pottery Neolithic' period. This observation served as a basis for recognizing a cultural gap of unknown duration. Originally, this cultural change was seen as due to a climatic crisis which caused the depopulation of the southern Levant (Perrot 1968). Subsequent field research demonstrated that the cultural gap reflects a major shift in settlement pattern, which is also evidenced in the northern Levant (Akkermans *et al.* 1983; Akkermans & Duistermaat 1996; Özdogan & Basgelen 1999), as well as on the Anatolian plateau.

Other proposals for explaining the collapse of the PPNB were derived from comparisons with the contemporary ecological hardships caused by the Industrial Revolution and the ensuing rapid land development and population increase during the nineteenth and twentieth centuries. For example, the change within the sequence of 'Ain Ghazal was interpreted as occurring as a result of over-exploitation of pastures and tree felling (Rollefson 1990; Rollefson & Köhler-Rollefson 1989; Rollefson *et al.* 1992). To expect that the same processes took place in both Anatolia and the Levant is to employ the same cause across every ecological belt within the entire region.

Another perspective views the collapse as motivated by societal over-exploitation among powerfully unequal villages. Unfortunately we have no conspicuous archaeological evidence for the presence of a particular 'Big Man' or other kind of chief. In addition, in spite of the intricate exchange systems, there is as yet no material expression of the enslaving of smaller communities by larger, richer ones. This is not to say that there is no evidence of social ranking. Clear signs within communities are seen in the finds of isolated stamps, signifying the existence of personal property. Perhaps future excavations will record the presence of slaves, a known phenomenon from the sedentary villages of the north-west coast of North America (Ames 1995).

Today, as the image of the PPNB civilization is more developed than before, its collapse throughout the entire region should, as mentioned in the introductory comments, lead us to examine the possibility that an abrupt climatic change was responsible for the rapid worsening of the environmental conditions. It does seem that the climatic crisis around 8,400–8,200 BP as recorded in the ice cores, was the culprit. The impact of the change is reflected in various pollen cores in Greece (Rossignol-Strick 1995), in Anatolia (van Zeist & Bottema 1991), and the Levant (Baruch & Bottema 1999). Under the new circumstances, a complex society that subsisted on farming and herding, in which the demands of better-off individuals (or families) drove the flow of foreign commodities, could not continue to accumulate surplus. The shift in the pattern of seasonal precipitation imposed the need for a search for pastures further away and resulted in lower yields of summer harvests. Finally, the economic deterioration resulted in a societal change expressed in the disappearance of previously large villages and the establishment of smaller villages or hamlets. The new conditions probably enhanced the reliance on the more flexible subsistence strategy of pastoral nomads. However, a more detailed discussion of the cultural changes remains beyond the scope of this chapter.

FINAL COMMENT

Describing and interpreting social history through the interpretation of archaeological documents is a notoriously difficult task and the published results are open to criticism. Each interpreter employs his or her imagination to describe how things happened in the past. Personally I prefer to adhere to the archaeological observations and minimize the amount and scope of interpretative speculations. However, even the most cautious interpretation is a subject for debate. The lack of written documents is often considered as hampering the sound reconstruction of past social forms and institutions. Similarities between the Neolithic societies and their cultural expressions and the aspects of Mesopotamian societies as reflected in the literary sources indicate that this is an as yet untapped source for additional interpretation. However, as we move deeper into the past, we are left solely with the archaeological remains. Hence, viewing the Late Palaeolithic human groups as simple foragers seems the most parsimonious interpretation. The cyclical shifts from egalitarian to non-egalitarian societies, as expressed in the archaeological sequence of the Late Palaeolithic to the early Natufian, and then the Late Natufian to the Early Neolithic (PPNA) villages, seem to repeat in the Near East after the collapse of the PPNB civilization. The current challenge, as previously stated, would be to identify archaeologically the more particular social institutions such as shamans, leaders, or headmen, the presence of early priesthood, the economic elite, and the like. This chapter thus joins other papers (as cited throughout) in contributing to efforts to reconstruct the social history of the Near East prior to the emergence of chiefdoms and states.

Note. I am grateful to W.G. Runciman for the invitation to take part in an interesting symposium, and to A. Belfer-Cohen (Hebrew University) for constant discussions on prehistoric social issues and useful comments on the current chapter. The text of this chapter benefited from an earlier one contributed to a volume on Cypriot archaeology edited by S. Swiny. J. Dickinson improved the original text. Any shortcomings remain my own.

REFERENCES

AKKERMANS, P.A., BOERMA, J.A.K., CLASON, A.T., HILL, S.G., LOHOF, E., MEIKLEJOHN, C., le MIÈRE, M., MOLGAT, G.M.F., ROODENBERG, J.J., WATERBOLK-VAN ROOYEN, W. & VAN ZEIST, W. 1983: Bouqras revisited: preliminary report on a project in Eastern Syria. *Proceedings of the Prehistoric Society* 49, 335–72.

AKKERMANS, P.M.M.G. & DUISTERMAAT, K. 1996: Of storage and nomads. The sealings from Late Neolithic Sabi Abyad, Syria. Commentaires de R. Bernbeck, S. Cleuziou, M. Frangipane, A. Le Brun, H. Nissen, H.T. Wright et réponse des auteurs. *Paléorient* 22, 17–44.

ALLEY, R.B., MAYEWSKI, P.A., SOWERS, T., STUIVER, M., TAYLOR, K.C. & Clark, P.U. 1997: Holocene climatic instability: A prominent, widespread event 8200 yr ago. *Geology* 25, 483–6.

ALLEY, R.B., MEESE, D.A., SHUMAN, C.A., GOW, A.J., TAYLOR, K.C., GROOTES, P.M., WHITE, J.W.C., RAM, M., WADDINGTON, E.D., MAYEWSKI, P.A. & ZIELINSKI, G.A. 1993: Abrupt increase in Greenland snow accumulation at the end of the Younger Dryas event. *Nature* 362, 527–9.

AMES, K.M. 1991: Sedentism: a temporal shift or a transitional change in hunter-gatherer mobility patterns? In Gregg, S.A. (ed.), *Between Bands and States* (Carbondale, IL, Center for Archaeological Investigations, Southern Illinois University), 108–34.

AMES, K.M. 1995: Chiefly power and household production on the northwest Coast. In Price, T.D. & Feinman, G.M. (eds.), *Foundations of Social Inequality* (New York, Plenum Press), 155–88.

AMIRAN, R. 1962: Myths of the creation of man and the Jericho statues. *Bulletin of the American Schools of Oriental Research* 167, 23–5.

ANDERSON, P.C. 1998: History of harvesting and threshing techniques for cereals in the prehistoric Near East. In Damania, A.B., Valkoun, J., Willcox, G. & Qualset, C.O. (eds.), *The Origins of Agriculture and Crop Domestication* (Aleppo, Syria, ICARDA), 145–59.

ARNOLD, J.E. (ed.) 1996: *Emergent Complexity: the Evolution of Intermediate Societies* (Ann Arbor, University of Michigan).

AURENCHE, O. & KOZLOWSKI, S.K. 1999: *La naissance du néolithique au Proche Orient ou le paradis perdu* (Paris, Editions Errance).

BAR-MATHEWS, M., AYALON, A. & KAUFMAN, A. 1997: Late Quaternary paleoclimate in the eastern Mediterranean region from stable isotope analysis of speleothems at Soreq Cave, Israel. *Quaternary Research* 47, 155–68.

BAR-MATHEWS, M., AYALON, A., KAUFMAN, A. & WASSERBURG, G.J. 1999: The eastern Mediterranean paleoclimate as a reflection of regional events: Soreq cave, Israel. *Earth and Planetary Science Letters* 166, 85–95.

BAR-YOSEF, D.E. 1989: Late Paleolithic and Neolithic marine shells in the southern Levant as cultural markers. In Hayes, C.F. (ed.), *Shell Bead Conference* (Rochester, NY, Rochester Museum and Science Center), 169–74.

BAR-YOSEF, D.E. 1991: Changes in the selection of marine shells from the Natufian to the Neolithic. In Bar-Yosef, O. & Valla, F.R. (eds.), *The Natufian Culture in the Levant* (Ann Arbor, International Monographs in Prehistory), 629–36.

BAR-YOSEF MAYER, D.E. 1997: Neolithic shell bead production in Sinai (as950097). *Journal of Archaeological Science* 24, 97–112.

BAR-YOSEF, O. 1981: The 'Pre-Pottery Neolithic' period in the southern Levant. In Cauvin, J. & Sanlaville, P. (eds.), *Préhistoire du Levant* (Paris, CNRS), 551–70.

BAR-YOSEF, O. 1984: Seasonality among Neolithic hunter-gatherers in southern Sinai. In Clutton-Brock, J. & Grigson, C. (eds.), *Animals and Archaeology, 3. Early Herders and their Flocks* (Oxford, British Archaeological Reports International Series 202), 145–60.

BAR-YOSEF, O. 1986: The walls of Jericho: an alternative interpretation. *Current Anthropology* 27, 157–62.

BAR-YOSEF, O. 1987: The Late Pleistocene in the Levant. In Soffer, O. (ed.), *The Pleistocene Old World: Regional Perspectives* (New York, Plenum Press), 219–36.

BAR-YOSEF, O. 1996: The impact of Late Pleistocene–Early Holocene climatic changes on humans in southwest Asia. In Straus, L.G., Eriksen, B.V., Erlandson, J.M. & Yesner, D.R. (eds.), *Humans at the End of the Ice Age: the Archaeology of the Pleistocene–Holocene Transition* (New York, Plenum Press), 61–76.

BAR-YOSEF, O. 1997a: Late Pleistocene lithic traditions in the Near East and their expression in Early Neolithic assemblages. In Kozlowski, S.K. & Gebel, H.G.K. (eds.), *Neolithic Chipped Stone Industries of the Fertile Crescent and their Contemporaries in Adjacent Regions: Proceedings of the Second Workshop on PPN Chipped Lithic Industries, Warsaw 1995* (Berlin, Ex Oriente), 207–16.

BAR-YOSEF, O. 1997b: Symbolic expressions in later prehistory of the Levant: why are they so few? In Conkey, M.W., Soffer, O., Stratmann, D. & Jablonski, N.G. (eds.), *Beyond Art: Pleistocene Image and Symbol* (San Francisco, Memoirs of the California Academy of Science), 161–87.

BAR-YOSEF, O. 1998: The Natufian Culture in the Levant — threshold to the origins of agriculture. *Evolutionary Anthropology* 6, 159–77.

BAR-YOSEF, O. & ALON, D. 1988: Excavations in the Nahal Hemar Cave. *Atiqot* 18, 1–30.

BAR-YOSEF, O. & BELFER-COHEN, A. 1989a: The Levantine 'PPNB' interaction sphere. In Hershkovitz, I. (ed.), *People and Culture in Change* (Oxford, British Archaeological Reports International Series 508(i)), 59–72.

BAR-YOSEF, O. & BELFER-COHEN, A. 1989b: The origins of sedentism and farming communities in the Levant. *Journal of World Prehistory* 3, 447–98.

BAR-YOSEF, O. & BELFER-COHEN, A. 1991: From sedentary hunter-gatherers to territorial farmers in the Levant. In Gregg, S.A. (ed.), *Between Bands and States* (Carbondale, IL, Center for Archaeological Investigations, Southern Illinois University), 181–202.

BAR-YOSEF, O. & BELFER-COHEN, A. 1992: From foraging to farming in the Mediterranean Levant. In Gebauer, A.B. & Price, T.D. (eds.), *Transitions to Agriculture in Prehistory* (Madison, Prehistory Press), 21–48.

BAR-YOSEF, O. & BELFER-COHEN, A. 1998: Natufian imagery in perspective. *Rivista di Scienze Prehistoriche* 49, 247–63.

BAR-YOSEF, O. & BELFER-COHEN, A. 1999: Encoding information: unique Natufian objects from Hayonim Cave, western Galilee, Israel. *Antiquity* 73, 402–10.

BAR-YOSEF, O. & GOPHER, A. (eds.) 1997: *An Early Neolithic Village in the Jordan Valley, Part I: The Archaeology of Netiv Hagdud* (Cambridge, MA, Peabody Museum of Archaeology and Ethnology, Harvard University).

BAR-YOSEF, O. & MEADOW, R.H. 1995: The origins of agriculture in the Near East. In Price, T.D. & GEBAUER, A.B. (eds.), *Last Hunters, First Farmers: New Perspectives on the Prehistoric Transition to Agriculture* (Santa Fe, School of American Research Press), 39–94.

BARUCH, U. 1994: The late Quaternary pollen record of the Near East. In Bar-Yosef, O. & Kra, R. (eds.), *Late Quaternary Chronology and Paleoclimates of the Eastern Mediterranean* (Tucson and Cambridge, MA, Peabody Museum of Archaeology and Ethnology, Harvard University), 103–20.

BARUCH, U. & BOTTEMA, S. 1999: A new pollen diagram from Lake Hula: vegetational, climatic, and anthropologenic implications. In Kawanabe, H., Coulter, G.W. & Roosevelt, A.C. (eds.), *Ancient Lakes: Their Cultural and Biological Diversity* (Belgium, Kenobe Productions), 75–86.

BELFER-COHEN, A. 1988: The Natufian graveyard in Hayonim Cave. *Paléorient* 14, 297–308.

BELFER-COHEN, A. 1991a: Art items from Layer B, Hayonim Cave: a case study of art in a Natufian context. In Bar-Yosef, O. & Valla, F.R. (eds.), *The Natufian Culture in the Levant* (Ann Arbor, International Monographs in Prehistory), 569–88.

BELFER-COHEN, A. 1991b: The Natufian in the Levant. *Annual Review of Anthropology* 20, 167–86.

BELFER-COHEN, A. 1995: Rethinking social stratification in the Natufian Culture: the evidence from burials. In Campbell, S. & Green, A. (eds.), *The Archaeology of Death in the Ancient Near East* (Oxford, Oxbow Monograph 51), 9–16.

BELFER-COHEN, A. & BAR-YOSEF, O. 2000: Early sedentism in the Near East: a bumpy ride to village life. In Kuijt, I. (ed.) *Life in Neolithic Farming Communities: Social Organization, Identity, and Differentiation* (New York, Plenum Press), 19–37.

BELFER-COHEN, A., SCHEPARTZ, L. & ARENSBURG, B. 1991: New biological data for the Natufian populations in Israel. In Bar-Yosef, O. & Valla, F.R. (eds.), *The Natufian Culture in the Levant* (Ann Arbor, International Monographs in Prehistory), 411–24.

BENTLEY, G.R., JASIENSKA, G. & GOLDBERG, T. 1993: Is the fertility of agriculturalists higher than that of nonagriculturalists? *Current Anthropology* 34, 778–85.

BINFORD, L.R. 1980: Willow smoke and dogs' tails: hunter-gatherer settlement systems and archaeological site formation. *American Antiquity* 45, 4–20.

BORZIYAK, I.A. 1993: Subsistence practices of Late Paleolithic groups along the Dnestr River and its tributaries. In Soffer, O. & Praslov, N.D. (eds.), *From Kostenki to Clovis: Upper Paleolithic—Paleo-Indian Adaptations* (New York, Plenum Press), 67–84.

BROECKER, W.S. 1999: What if the conveyor were to shut down? Reflections on a possible outcome of the great global experiment. *GSA Today* 9, 1–7.

BYRD, B.F. 1994a: Late Quaternary hunter-gatherer complexes in the Levant between 20,000 and 10,000 BP. In Bar-Yosef, O. & Kra, R. (eds.), *Late Quaternary Chronology and Paleoclimates of the Eastern Mediterranean* (Tucson and Cambridge, MA, Peabody Museum of Archaeology and Ethnology, Harvard University), 205–26.

BYRD, B.F. 1994b: Public and private, domestic and corporate: the emergence of the southwest Asian village. *American Antiquity* 59, 639–66.

BYRD, B.F. 1998: Spanning the gap between the Upper Paleolithic and the Natufian: the Early and Middle Epipaleolithic. In Henry, D.O. (ed.), *The Prehistoric Archaeology of Jordan* (Oxford, Archaeopress), 64–82.

BYRD, B.F. 2000: Households in transition: Neolithic social organization within southwest Asia. In Kuijt, I. (ed.), *Life in Neolithic Farming Communities: Social Organization, Identity, and Differentiation* (New York, Plenum Press), 63–98.

BYRD, B.F. & MONAHAN, C.M. 1995: Death, mortuary ritual, and Natufian social structure. *Journal of Anthropological Archaeology* 14, 251–87.

CAMERON, C.M. & TOMKA, S.A. 1993: *Abandonment of Settlements and Regions: Ethnoarchaeological and Archaeological Approaches* (Cambridge and New York, Cambridge University Press).

CAMPANA, D.V. 1989: *Natufian and Protoneolithic Bone Tools* (Oxford, British Archaeological Reports International Series 494).

CAPPERS, R.T.J., BOTTEMA, S. & WOLDRING, H. 1998: Problems in correlating pollen diagrams of the Near East: a preliminary report. In Damania, A.B., Valkoun, J., Willcox, G. & Qualset, C.O. (eds.), *The Origins of Agriculture and Crop Domestication* (Aleppo, Syria, ICARDA), 160–9.

CAUVIN, J. 1977: Les fouilles de Mureybet (1971–1974) et leur signification pour les origines de la sédentarisation au Proche-Orient. *Annual of the American Schools of Oriental Research* 44, 19–48.

CAUVIN, J. 1978: *Les premiers villages de Syrie-Palestine de IXeme au VIIeme millenaire avant J.C.* (Lyon, Maison de l'Orient).

CAUVIN, J. 1985: La question du 'Matriarcat Préhistorique' et le rôle de la femme dans la Préhistoire. In Vérilhac, A.M. (ed.), *La femme dans le monde Mediterranéen* (Lyon, Maison de l'Orient), 7–18.

CAUVIN, J. 1997: *Naissance des divinités, naissance de l'agriculture* (Paris, CNRS).

CAUVIN, J., AURENCHE, O., CAUVIN, M.-C. & BALKAN-ATLI, N. 1999: The pre-pottery site of Cafer Höyük. In Özdogan, M. & Basgelen, N. (eds.), *Neolithic in Turkey: Cradle of Civilization. New Discoveries* (Istanbul, Arkeoloji ve Sanat Yayinlari), 87–104.

CAUVIN, M.-C. & STORDEUR, D. 1978: *Les outillages lithiques et osseux de Mureybet, Syrie* (Paris, CNRS).

COLLEDGE, S. 1998: Identifying pre-domestication cultivation using multivariate analysis. In Damania, A.B., Valkoun, J., Willcox, G. & Qualset, C.O. (eds.), *The Origins of Agriculture and Crop Domestication* (Aleppo, Syria, ICARDA), 121–31.

COUPLAND, G. 1996: This old house: cultural complexity and household stability on the northern north-west coast of North America. In Arnold, J.E. (ed.), *Emergent Complexity: the Evolution of Intermediate Societies* (Ann Arbor, International Monographs in Prehistory), 74–90.

DANIN, A. 1988: Flora and vegetation of Israel and adjacent areas. In Yom-Tov, Y. & Tchernov, E. (eds.), *The Zoogeography of Israel* (Dordrecht, Dr. W. Junk Publishers), 129–58.

DAVIS, S.J.M. 1982: Climatic change and the advent of domestication of ruminant artiodactyls in the Late Pleistocene–Holocene period in the Israel region. *Paléorient* 8, 5–16.

DAVIS, S.J.M. & VALLA, F. 1978: Evidences for the domestication of the dog in the Natufian of Israel 12,000 years ago. *Nature* 276, 608–10.

DOBRES, M. & HOFFMAN, C. 1994: Social agency and the dynamics of prehistoric technology. *Journal of Archaeological Method and Theory* 1, 211–58.

EARLE, T. 1997: *How Chiefs Come to Power: the Political Economy in Prehistory* (Stanford, Stanford University Press).

FELLNER, R.O. 1995: *Cultural Change and the Epipalaeolithic of Palestine* (Oxford, Tempus Reparatum).

FLANNERY, K.V. 1972: The origins of the village as a settlement type in Mesoamerica and the Near East: a comparative study. In Ucko, P.J., Trigham, R. & Dimbleby, G.W. (eds.), *Man, Settlement and Urbanism* (London, Duckworth), 23–53.

FRUMKIN, A., FORD, D.C. & SCHWARCZ, H.P. 1999: Continental oxygen isotopic record of the last 170,000 years in Jerusalem. *Quaternary Research* 51, 317–27.

GALILI, E. & NIR, Y. 1993: The submerged Pre-Pottery Neolithic water well of Atlit-Yam, northern Israel, and its paleoenvironmental implications. *The Holocene* 3, 265–70.

GARFINKEL, Y. 1993: The Yarmukian culture in Israel. *Paléorient* 19, 115–34.

GARFINKEL, Y. 1994: Ritual burial of cultic objects: the earliest evidence. *Cambridge Archaeological Journal* 4, 159–88.

GARFINKEL, Y. 1999: Radiometric dates from eighth millennium BP Israel. *BASOR* 315, 1–13.

GARRARD, A.N. 1998: Palaeolithic and Neolithic survey at a south-eastern 'Gateway' to Turkey. In Matthews, R. (ed.), *Ancient Anatolia: Fifty Years' Work by the British Institute of Archaeology at Ankara* (Ankara, The British Institute of Archaeology at Ankara), 7–16.

GARRARD, A.N. 1999: Charting the emergence of cereal and pulse domestication in south west Asia. *Environmental Archaeology* 4, 67–86.

GARRARD, A.N., BAIRD, D. & BYRD, B.F. 1994: The chronological basis and significance of the Late Paleolithic and Neolithic sequence in the Azraq basin, Jordan. In Bar-Yosef, O. & Kra, R. (eds.), *Late Quaternary Chronology and Paleoclimates of the Eastern Mediterranean* (Tucson and Cambridge, MA, Peabody Museum of Archaeology and Ethnology, Harvard University), 177–200.

GARRARD, A.N., BETTS, A., BYRD, B. & HUNT, C. 1988: Summary of palaeoenvironmental and prehistoric investigations in the Azraq basin. In Garrard, A.N. & Gebel, H.G. (eds.), *The Prehistory of Jordan* (Oxford, British Archaeological Reports International Series S396 (i)), 311–37.

GARRARD, A.N., COLLEDGE, S. & MARTIN, L. 1996: The emergence of crop cultivation and caprine herding in the 'Marginal Zone' of the southern Levant. In Harris, D. (ed.), *The Origins and Spread of Agriculture and Pastoralism in Eurasia* (London, UCL Press), 204–26.

GLANTZ, M.H. (ed.) 1987: *Drought and Hunger in Africa* (Cambridge, Cambridge University Press).

GLANTZ, M.H. (ed.) 1994: *Drought Follows the Plow: Cultivating Marginal Areas* (Cambridge, Cambridge University Press).

GLANTZ, M.H., STREETS, D.G., STEWART, T.R., BHATTI, N., MOORE, C.M. & ROSA, C.H. 1998: *Exploring the Concept of Climate Surprises: a Review of the Literature on the Concept of Surprise and How it is Related to Climate Change* (Argonne, IL, US Department of Energy, Office of Energy Research).

GOPHER, A. 1994a: *Arrowheads of the Neolithic Levant: a Seriation Analysis* (Winona Lake, IN, Eisenbrauns).

GOPHER, A. 1994b: Southern-central Levant PPN cultural sequences: time–space systematics through typological and stylistic approaches. In Gebel, H.G. & Kozlowski, S.K. (eds.), *Neolithic Chipped Stone Industries of the Fertile Crescent: Proceedings of the First Workshop on PPN Chipped Lithic Industries* (Berlin, Ex Oriente), 387–92.

GOPHER, A. & GOPHNA, R. 1993: Cultures of the eighth and seventh millennium BP in southern Levant: a review for the 1990s. *Journal of World Prehistory* 7, 297–351.

GORING-MORRIS, A.N. 1987: *At the Edge: Terminal Pleistocene Hunter-Gatherers in the Negev and Sinai* (Oxford, British Archaeological Reports 361).

GORING-MORRIS, A.N. 1991: The Harifian of the southern Levant. In Bar-Yosef, O. & VALLA, F.R. (eds.), *The Natufian Culture in the Levant* (Ann Arbor, International Monographs in Prehistory), 173–216.

GORING-MORRIS, A.N. 1995: Complex hunter-gatherers at the end of the Palaeolithic (20,000–10,000 BP). In Levy, T.E. (ed.), *The Archaeology of Society in the Holy Land* (London, Leicester University Press), 141–68.

GORING-MORRIS, A.N. & BELFER-COHEN, A. 1997: The articulation of cultural processes and late quaternary environmental changes in Cisjordan. *Paléorient* 23, 71–94.

GORING-MORRIS, A.N., GOREN, Y., HORWITZ, L.K., BAR-YOSEF, D. & HERSHKOVITZ, I. 1995: Investigations at an early Neolithic settlement in the Lower Galilee: results of the 1991 season at Kefar HaHoresh. *Atiqot* 27, 37–62.

GORING-MORRIS, A.N., MARDER, O., DAVIDZON, A. & IBRAHIM, F. 1998: Putting Humpty together again: preliminary observations on refitting studies in the eastern Mediterranean. In Milliken, S. (ed.), *The Organization of Lithic Technology in Late Glacial and Early Postglacial Europe* (Oxford, British Archaeological Reports International Series 700), 149–82.

HAIM, A. & TCHERNOV, E. 1974: The distribution of myomorph rodents in the Sinai peninsula. *Mammalia* 38, 201–23.

HASSAN, F.A. 1997: Holocene palaeoclimates of Africa. *African Archaeological Review* 14, 213–30.

HAUPTMANN, H. 1999: The Urfa region. In Özdogan, M. & Basgelen, N. (eds.), *Neolithic in Turkey: Cradle of Civilization. New Discoveries* (Istanbul, Arkeoloji ve Sanat Yayinlari), 65–86.

HAYDEN, B. 1995: Pathways to power: principles for creating socioeconomic inequalities. In Price, T.D. & Feinman, G.M. (eds.), *Foundations of Social Inequality* (New York, Plenum Press), 15–86.

HENRY, D.O. 1989: *From Foraging to Agriculture: the Levant at the End of the Ice Age* (Philadelphia, University of Pennsylvania Press).

HENRY, D.O. 1991: Foraging, sedentism, and adaptive vigor in the Natufian: rethinking the linkages. In Clark, G.A. (ed.), *Perspectives on the Past: Theoretical Biases in Mediterranean Hunter-Gatherer Research* (Philadelphia, University of Pennsylvania Press), 353–70.

HENRY, D.O. 1995: *Prehistoric Cultural Ecology and Evolution* (New York, Plenum Press).

HENRY, D.O. 1997: Prehistoric human ecology in the southern Levant east of the Rift from 20 000–6 000 BP. *Paléorient* 23, 107–20.

HERSHKOVITZ, I., BAR-YOSEF, O. & ARENSBURG, B. 1994: The Pre-Pottery Neolithic populations of south Sinai and their relations to other circum-Mediterranean groups: anthropological study. *Paléorient* 20, 59–84.

HESSE, B. 1984: These are our goats: the origins of herding in west central Iran. In Clutton-Brock, J. & Grigson, C. (eds.), *Animal and Archaeology, 3. Early Herders and their Flocks* (Oxford, British Archaeological Reports International Series 202), 243–64.

HILLMAN, G. 1996: Late Pleistocene changes in wild plant-foods available to hunter-gatherers of the Northern Fertile Crescent: possible preludes to cereal cultivation. In Harris, D. (ed.), *The Origins and Spread of Agriculture and Pastoralism in Eurasia* (London, UCL Press), 159–203.

HILLMAN, G.C., COLLEDGE, S. & HARRIS, D.R. 1989: Plant food economy during the epi-Palaeolithic period at Tell Abu Hureyra, Syria: dietary diversity, seasonality and modes of exploitation. In Harris, D.R. & Hillman, G.C. (eds.), *Foraging and Farming: the Evolution of Plant Exploitation* (London, Unwin Hyman), 240–66.

HODDER, I. 1999: Renewed work at Çatalhöyük. In Özdogan, M. & Basgelen, N. (eds.), *Neolithic in Turkey: Cradle of Civilization. New Discoveries* (Istanbul, Arkeoloji ve Sanat Yayinlari), 157–64.

HOLE, F. 1994: Interregional aspects of the Khuzestan Aceramic–Early Pottery Neolithic sequence. In Gebel, H.G. & Kozlowski, S. K. (eds.), *Neolithic Chipped Stone Industries of the Fertile Crescent: Proceedings of the First Workshop on PPN Chipped Lithic Industries* (Berlin, Ex Oriente), 101–16.

JELINEK, J. 1999: Behaviour and survival strategy in Moravian early Gravettians: mammoth hunters or scavengers. In Ullrich, H. (ed.), *Hominid Evolution: Lifestyles and Survival Strategies* (Gelsenkirchen, Edition Archaea), 457–79.

JOHNSON, A.W. & EARLE, T. 1987: *The Evolution of Human Societies: From Foraging Group to Agrarian State* (Stanford, Stanford University Press).

JOHNSON, G.A. 1987: The changing organization of Uruk administration on the Susiana Plain. In Hole, F. (ed.), *The Archaeology of Western Iran: Settlement and Society from Prehistory to the Islamic Conquest* (Washington, DC, Smithsonian Institution Press), 107–39.

KEELEY, L.H. 1988: Hunter-gatherer economic complexity and 'population pressure': a cross-cultural analysis. *Journal of Anthropological Archaeology* 7, 373–411.

KEELEY, L.H. 1996: *War Before Civilization* (New York, Oxford University Press).

KELLY, R. 1995: *The Foraging Spectrum: Diversity in Hunter-Gatherer Lifeways* (Washington, DC, Smithsonian Institution Press).

KELLY, R.L. 1991: Sedentism, socio-political inequality, and resource fluctuations. In Gregg, S.A. (ed.), *Between Bands and States* (Carbondale, IL, Center for Archaeological Investigations, Southern Illinois University Press), 135–58.

KENYON, K. 1957: *Digging Up Jericho* (London, Benn).

KENYON, K. & HOLLAND, T. 1981: *Excavations at Jericho, Vol. III: the Architecture and Stratigraphy of the Tell* (London, British School of Archaeology in Jerusalem).

KIRKBRIDE, D. 1968: Beidha: Early Neolithic village life south of the Dead Sea. *Antiquity* 42, 263–74.

KISLEV, M.E. 1989: Pre-domesticated cereals in the Pre-Pottery Neolithic A Period. In Hershkovitz, I. (ed.), *People and Culture Change* (Oxford, British Archaeological Reports International Series 508i), 147–52.

KISLEV, M.E. 1997: Early agriculture and paleoecology of Netiv Hagdud. In Bar-Yosef, O. & Gopher, A. (eds.), *An Early Neolithic Village in the Jordan Valley Part I: the Archaeology of Netiv Hagdud* (Cambridge, MA, Peabody Museum of Archaeology and Ethnology, Harvard University), 209–36.

KISLEV, M.E., NADEL, D. & CARMI, I. 1992: Epi-Palaeolithic (19,000 BP) cereal and fruit diet at Ohalo II, Sea of Galilee, Israel. *Review of Palaeobotany and Palynology* 71, 161–6.

KOSSE, K. 1994: The evolution of large, complex groups: a hypothesis. *Journal of Anthropological Archaeology* 13, 35–50.

KOZLOWSKI, S.K. 1994: Chipped Neolithic industries at the eastern wing of the Fertile Crescent. In Gebel, H.G. & Kozlowski, S.K. (eds.), *Neolithic Chipped Stone Industries of the Fertile Crescent: Proceedings of the First Workshop on PPN Chipped Lithic Industries* (Berlin, Ex Oriente), 143–72.

KOZLOWSKI, S.K. 1999: *The Eastern Wing of the Fertile Crescent: Late Prehistory of Greater Mesopotamian Lithic Industries* (Oxford, Archaeopress).

KOZLOWSKI, S.K. & GEBEL, H.G.K. (eds.) 1996: *Neolithic Chipped Stone Industries of the Fertile Crescent and their Contemporaries in Adjacent Regions: Proceedings of the Second Workshop on PPN Chipped Lithic Industries, Warsaw 1995* (Berlin, Ex Oriente).

KRAMER, C. 1983: Spatial organization in contemporary southwest Asian villages. In Young, T.C., Jr., Smith, P.E.L. & Mortensen, P. (eds.), *The Hilly Flanks and Beyond* (Chicago, The Oriental Institute), 347–68.

KUIJT, I. 1996: Negotiating equality through ritual: a consideration of Late Natufian and Prepottery Neolithic A period mortuary practices. *Journal of Anthropological Archaeology* 15, 313–36.

KUIJT, I. (ed.) 2000a: *Life in Neolithic Farming Communities: Social Organization, Identity, and Differentiation* (New York, Plenum Press).

KUIJT, I. 2000b: People and space in early agricultural villages: exploring daily lives, community size, and architecture in the late Pre-Pottery Neolithic. *Journal of Anthropological Archaeology* 19, 75–102.

LAMBECK, K. & BARD, E. 2000: Sea-level change along the French Mediterranean coast for the past 30 000 years. *Earth and Planetary Science Letters* 175, 203–22.

LEGGE, T. 1996: The beginning of caprine domestication in southwest Asia. In Harris, D. (ed.), *The Origins and Spread of Agriculture and Pastoralism in Eurasia* (London, UCL Press), 238–62.

LEMCKE, G. & STURM, M. 1997: ∂18O and trace element measurements as proxy for the reconstructions of climate changes at Lake Van (Turkey): preliminary results. In Dalfes, H.N., Kukla, G. & Weiss, H. (eds.), *Third Millennium BC Climate Change and Old World Collapse* (Berlin, Springer-Verlag), 653–78.

LEMONNIER, P. 1992: *Elements for an Anthropology of Technology* (Ann Arbor, University of Michigan).

LÉVI-STRAUSS, C. 1983: *The Way of the Masks* (London, Jonathan Cape).

McCORRISTON, J. & HOLE, F. 1991: The ecology of seasonal stress and the origins of agriculture in the Near East. *American Anthropologist* 93, 46–94.

MARSHACK, A. 1997: Paleolithic image making and symboling in Europe and the Middle East: a comparative review. In Conkey, M., Soffer, O., Stratmann, D. & Jablonski, N.G. (eds.), *Beyond Art: Pleistocene Image and Symbol* (San Francisco, Memoirs of California Academy of Sciences), 53–91.

MARX, E. 1977: Communal and individual pilgrimage: the region of saints' tombs in south Sinai. In Werbner, R.P. (ed.), *Regional Cults* (The Hague, Mouton), 29–51.

MAYEWSKI, P.A. & BENDER, M. 1995: The GISP2 ice core record — paleoclimate highlights. *Reviews of Geophysics, Supplement* July, 1287–96.

MELLAART, J. 1967: *Çatalhöyük, a Neolithic Town in Anatolia* (London, Thames & Hudson).

MESHEL, Z. 1974: New data about the 'Desert Kites'. *Tel Aviv* 1, 129–43.

MOORE, A.M.T. 1989: The transition from foraging to farming in southwest Asia: present problems and future directions. In Harris, D.R. & Hillman, G.C. (eds.), *Foraging and Farming: the Evolution of Plant Exploitation* (London, Unwin Hyman), 620–31.

MOORE, A.M.T. & HILLMAN, G.C. 1992: The Pleistocene to Holocene transition and human economy in southwest Asia: the impact of the Younger Dryas. *American Antiquity* 57, 482–94.

O'BRIEN, S.R., MAYEWSKI, P.A., MEEKER, L.D., MEESE, D.A., TWICKLER, M.S. & WHITLOW, S. I. 1995: Complexity of Holocene climate as reconstructed from a Greenland ice core. *Science* 270, 1962–4.

OLIVER, P. 1971: *Shelter in Africa* (New York, Praeger).

OTTERBEIN, K. 1997: The origins of war. *Critical Review* 2, 251–77.

ÖZDOGAN, A. 1999: Çayönü. In Özdogan, M. & Basgelen, N. (eds.), *Neolithic in Turkey: Cradle of Civilization. New Discoveries* (Istanbul, Arkeoloji ve Sanat Yayinlari), 35–64.

ÖZDOGAN, M. & BASGELEN, N. (eds.) 1999: *Neolithic in Turkey: Cradle of Civilization. New Discoveries* (3 vols., Istanbul, Turkey, Arkeoloji ve Sanat Yayinlari).

ÖZDOGAN, M. & ÖZDOGAN, A. 1998: Buildings of cult and the cult of buildings. In Arsebük, G., Mellink, M.J. & Schirmer, W. (eds.), *Light on Top of the Black Hill. Studies presented to Halet Çambel* (Istanbul, Ege Yayinlari), 581–601.

PERROT, J. 1966: Le gisement natoufien de Mallaha (Eynan), Israël. *L'Anthropologie* 70, 437–84.

PERROT, J. 1968: La préhistoire palestinienne. In *Supplément au Dictionnaire de la Bible 8* (Paris, Letouzey & Ané), cols. 286–446.

PRICE, T.D. & BROWN, J.A. 1985: *Prehistoric Hunter-Gatherers: the Emergence of Cultural Complexity* (Orlando, Academic Press).

PRICE, T.D. & GEBAUER, A.B. (eds.) 1995: *Last Hunters — First Farmers: New Perspectives on the Prehistoric Transition to Agriculture* (Santa Fe, School of American Research Press).

QUINTERO, L.A. & WILKE, P.J. 1995: Evolution and economic significance of naviform core-and-blade technology in the southern Levant. *Paléorient* 21, 17–33.

RABINOVICH, R. 1998: 'Patterns of Animal Exploitation and Subsistence in Israel During the Upper Palaeolithic and Epi-Palaeolithic (40,000–12,500 BP), Based upon Selected Case Studies.' PhD thesis, The Hebrew University, Jerusalem.

RAFFERTY, G.E. 1985: The archaeological record on sedentariness: recognition, development and implications. In Schiffer, M.B. (ed.), *Advances in Archaeological Method and Theory* (New York, Academic Press), 113–56.

RAPOPORT, A. 1969: *House Form and Culture* (Englewood Cliffs, NJ, Prentice-Hall).

REESE, D.S. 1991: Marine shells in the Levant: Upper Paleolithic, Epipaleolithic, and Neolithic. In Bar-Yosef, O. & Valla, F.R. (eds.), *The Natufian Culture in the Levant* (Ann Arbor, International Monographs in Prehistory), 613–28.

ROLLEFSON, G.O. 1983: Ritual and ceremony at Neolithic 'Ain Ghazal (Jordan). *Paléorient* 9, 29–38.

ROLLEFSON, G.O. 1990: The uses of plaster at Neolithic 'Ain Ghazal, Jordan. *Archeomaterials* 4, 33–54.

ROLLEFSON, G.O. & KÖHLER-ROLLEFSON, I. 1989: The collapse of Early Neolithic settlements in the southern Levant. In Hershkovitz, I. (ed.), *People and Culture in Change: Proceedings of the Second Symposium on Upper Palaeolithic, Mesolithic and Neolithic Populations of Europe and the Mediterranean Basin* (Oxford, British Archaeological Reports 58(i)), 73–89.

ROLLEFSON, G.O., SIMMONS, A.H. & KAFAFI, Z. 1992: Neolithic cultures at 'Ain Ghazal, Jordan. *Journal of Field Archaeology* 19, 443–70.

ROSENBERG, M. 1998: Cheating at musical chairs: territoriality and sedentism in an evolutionary context. *Current Anthropology* 39, 653–81.

ROSENBERG, M., NESBITT, R., REDDING, R.W. & PEASNALL, B.L. 1998: Hallan Çemi, pig husbandry, and post Pleistocene adaptations among the Taurus–Zagros Arc (Turkey). *Paléorient* 24, 25–41.

ROSSIGNOL-STRICK, M. 1995: Sea–land correlation of pollen records in the eastern Mediterranean for the glacial–interglacial transition: biostratigraphy versus radiometric timescale. *Quaternary Science Reviews* 14, 893–915.

ROSSIGNOL-STRICK, M. 1997: Paléoclimat de la Méditerranée orientale et de l'Asie du Sud-Ouest de 15 000 à 6 000 BP. *Paléorient* 23, 175–86.

SAGE, R.F. 1995: Was low atmospheric CO_2 during the Pleistocene a limiting factor for the origin of agriculture? *Global Change Biology* 1, 93–106.

SANLAVILLE, P. 1996: Changements climatiques dans la région levantine à la fin du Pléistocène supérieur et au début de l'Holocène. Leurs relations avec l'évolution des sociétés humaines. *Paléorient* 22, 7–30.

SANLAVILLE, P. 1997: Les changements dans l'environnement au Moyen-Orient de 20 000 BP à 6 000 BP. *Paléorient* 23, 249–62.

SCHMIDT, K. 1999: Boars, ducks, and foxes — the Urfa-Project 99. *Neo-Lithics* 3, 12–15.

SERGUIN, V.I. 1999: Zhilishcha na pamiatnikakh vostochnogo gravetta Russky Ravniny. In Amirkhanov, A.A. (ed.), *Vostochny Gravett* (Moskow, Nauchni Mir), 151–76.

SERVICE, E.R. 1962: *Primitive Social Organization: an Evolutionary Perspective* (New York, Random House).

SMITH, P.E.L. 1976: Reflections on four seasons of excavations at Tapeh Ganj Dareh. In Bagherzadeh, F. (ed.), *Proceedings of the IVth Annual Symposium on Archaeological Research in Iran* (Tehran, Iranian Centre for Archaeological Research), 11–22.

SOLECKI, R.L. 1981: *An Early Village Site at Zawi Chemi Shanidar* (Malibu, Undena Publications).

STARK, B. 1986: Origins of food production in the New World. In Melzer, D.J., Fowler, D.D. & Sabloff, J.A. (eds.), *American Archaeology Past and Future* (Washington, DC, Smithsonian Institution Press), 277–321.

STEKELIS, M. 1972: *The Yarmukian Culture* (Jerusalem, Magness Press, The Hebrew University).

STORDEUR, D., HELMER, D. & WILCOX, G. 1997: Jerf el Ahmar: un nouveau site de l'horizon PPNA sur le moyen Euphrate syrien. *Bulletin de la Société Préhistorique Française* 94, 282–5.

STUIVER, M., REIMER, P.J., BARD, E., BECK, J.W., BURR, G.S., HUGHEN, K.A., KROMER, B., McCORMAC, G., van der PLICHT, J. & SPURK, M. 1998: INTCAL98 radiocarbon Aage calibration, 24,000–0 cal BP. *Radiocarbon* 40, 1041–84.

SVOBODA, J., LOZEK, V. & VLCEK, E. 1996: *Hunters Between East and West: the Paleolithic of Moravia* (New York, Plenum Press).

TAYLOR, K.C., MAYEWSKI, P.A., ALLEY, R.B., BROOK, E.J., GOW, A.J., GROOTES, P.M., MEESE, D.A., SALTZMAN, E.S., SEVERINGHAUS, J.P., TWICKLER, M.S., WHITE, J.W.C., WHITLOW, S. & ZIELINSKI, G.A. 1997: The Holocene–Younger Dryas transition recorded at Summit, Greenland. *Science* 278, 825–7.

TCHERNOV, E. 1991a: Biological evidence for human sedentism in southwest Asia during the Natufian. In Bar-Yosef, O. & Valla, F.R. (eds.), *The Natufian Culture in the Levant* (Ann Arbor, International Monographs in Prehistory), 315–40.

TCHERNOV, E. 1991b: On mice and men: biological markers for long-term sedentism: a reply. *Paléorient* 17, 153–60.

TCHERNOV, E. 1993a: The effects of sedentism on the exploitation of the environment in the southern Levant. In Desse, J. & Audoin-Rouzeau, F. (eds.), *Exploitation des animaux sauvages à travers le temps* (Juan-les-Pins, APDCA), 137–59.

TCHERNOV, E. 1993b: From sedentism to domestication — a preliminary review for the southern Levant. In Clason, A., Payne, S. & Uerpmann, H.P. (eds.), *Skeletons in her Cupboard: Festschrift for Juliet Clutton-Brock* (Oxford, Oxbow Monograph 34), 189–233.

TCHERNOV, E. 1994: *An Early Neolithic Village in the Jordan Valley II: the Fauna of Netiv Hagdud* (Cambridge, MA, Peabody Museum of Archaeology and Ethnography, Harvard University).

TCHERNOV, E. & VALLA, F. 1997: Two new dogs, and other Natufian dogs, from the southern Levant. *Journal of Archaeological Science* 24, 65–95.

UNGER-HAMILTON, R. 1991: Natufian plant husbandry in the southern Levant and comparison with that of the Neolithic periods: the lithic perspective. In Bar-Yosef, O. & Valla, F.R.

(eds.), *The Natufian Culture in the Levant* (Ann Arbor, International Monographs in Prehistory), 483–520.

UPHAM, S. 1990: Decoupling the processes of political evolution. In Upham, S. (ed.), *The Evolution of Political Systems* (Cambridge, Cambridge University Press), 1–17.

VALLA, F.R. 1987: Les Natoufiens connaissaient-ils l'arc? In Stordeur, D. (ed.), *La main et l'outil: manches et emmanchements préhistoriques* (Lyon, Maison de l'Orient), 165–74.

VALLA, F.R. 1990: Le Natoufien: une autre façon de comprendre le monde? *Journal of the Israel Prehistoric Society* 23, 171–5.

VALLA, F.R. 1991: Les Natoufiens de Mallaha et l'espace. In Bar-Yosef, O. & Valla, F.R. (eds.), *The Natufian Culture in the Levant* (Ann Arbor, International Monographs in Prehistory), 111–22.

VALLA, F.R. 1995: The first settled societies — Natufian (12,500–10,200 BP). In Levy, T. (ed.), *The Archaeology of Society in the Holy Land* (London, Leicester University Press), 169–89.

VALLA, F.R., KHALAILY, H., SAMUELIAN, N., BOUCQUENTIN, F., DELAGE, C., VALENTIN, B., PLISSON, H., RABINOVICH, R. & BELFER-COHEN, A. 1998: Le natoufien final et les nouvelles fouilles à Mallaha. *Mitekufat Haeven, Journal of the Israel Prehistoric Society* 28, 105–76.

VAN ZEIST, W. & BAKKER-HERRES, J.A.H. 1986: Archaeobotanical studies in the Levant. III. Late Paleolithic Mureybet. *Palaeohistoria* 26, 171–99.

VAN ZEIST, W. & BOTTEMA, S. 1991: *Late Quaternary Vegetation of the Near East* (Wiesbaden, Dr Ludwig Reichert Verlag).

VIGNE, J.D., BUITENHUIS, H. & DAVIS, S. 1999: Les premiers pas de la domestication animale à l'Ouest de l'Euphrate: Chypre et l'Anatohe centrale. *Paléorient* 25, 49–62.

VOIGT, M.M. 1990: Reconstructing Neolithic societies and economies in the Middle East: an essay. *Archaeomaterials* 4, 1–14.

WATKINS, T., BAIRD, D. & BETTS, A. 1989: Qermez Dere and the early Aceramic Neolithic in N. Iraq. *Paléorient* 15, 19–24.

WEINSTEIN-EVRON, M. 1997: The Natufian use of el-Wad Cave, Mount Carmel, Israel. In Bonsall, C. & Tolan-Smith, C. (eds.), *The Human Use of Caves* (Oxford, British Archaeological Reports International Series 667), 155–66.

WEINSTEIN-EVRON, M., LANG, B., ILANI, S., STEINITZ, G. & KAUFMAN, D. 1995: K/AR dating as a means of sourcing Levantine Epipalaeolithic basalt implements. *Archaeometry* 37, 37–40.

WILKE, P.J. & QUINTERO, L.A. 1994: Naviform core-and-blade technology: assemblage character as determined by replicative experiments. In Gebel, H.G. & Kozlowski, S.K. (eds.), *Neolithic Chipped Stone Industries of the Fertile Crescent: Proceedings of the First Workshop on PPN Chipped Lithic Industries* (Berlin, Ex Oriente), 33–60.

WRIGHT, G.A. 1978: Social differentiation in the Early Natufian. In Redman, C.L., Berman, M.J., Curint, E.V., Langhorne, W.T.J., VersaggI, N.M. & Wanser, J.C. (eds.), *Social Archaeology: Beyond Subsistence and Dating* (New York, Academic Press), 201–33.

YAMADA, S. 2000: 'Development of the Neolithic: Lithic Use-Wear Analysis of Major Tool Types in the Southern Levant.' PhD dissertation, Harvard University, Cambridge, MA.

ZEDER, M.A. & HESSE, B. 2000: The initial domestication of goats (*Capra hircus*) in the Zagros Mountains 10,000 Years Ago. *Science* 287, 2254–7.

ZOHARY, D. 1992: Domestication of the Neolithic Near East crop assemblage. In Anderson, P. C. (ed.), *Préhistoire de l'agriculture* (Paris, CNRS), 81–6.

ZOLITSCHKA, B., HAVERKAMP, B. & NEGENDANK, J.F.W. 1992: Younger Dryas oscillation — varve dated microstratigraphic, palynological and palaeomagnetic records from Lake Holzmaar, Germany. In Bard, E. & Broecker, W.S. (eds.), *The Last Deglaciation: Absolute and Radiocarbon Chronologies* (Berlin, Springer-Verlag), 81–101.

Different Kinds of History:
On the Nature of Lives and Change in
Central Europe, *c*. 6000 to the
Second Millennium BC

ALASDAIR WHITTLE

LONG VIEWS OF DIFFERENCES BETWEEN THE NEAR EAST AND EUROPE

THE PAST WAS NOT JUST ONE other country, but many. In the ancient Near East, we know of individuals who would have understood the nature, if not always the causes, of inherited status, central control of stored resources, monumental architecture, markets, armies, and extravagant death rituals expressive of an ideology of personal and divinely sanctioned power. The early historical record of the area documents many such individuals already in the mid- to later third millennium BC, such as Ur-Nanshe and his son Akurgal of Lagash, Urnammu of Ur, founder of the Third Dynasty, or Sargon of Akkad, founder of the Akkadian dynasty, and his priestess daughter Enheduanna (Leick 1999). There were presumably predecessors, now undocumented as individuals, comparable to Narmer and the early pharaohs in Egypt (Kemp 1989), who would have lived through the key changes in political and economic centralization that constitute the emergence of urban states in the Near East around the end of the fourth millennium BC (Postgate 1992).

Things were never like this in prehistoric temperate (that is, non-Mediterranean) Europe. This difference has been recognized since before the days of Gordon Childe, who himself gave much attention to it and its explanation in the middle part of the twentieth century, and it has been repeated many times since (e.g. Sherratt 1982: 13; 1994, 1995). Many authors agree that it is not until as late as the mid-second millennium BC that small-scale chiefdoms appear in central Europe (e.g. S.J. Shennan 1986, 1993), coinciding in part with the appearance of a male warrior elite or aristocracy (Kristiansen 1998; Treherne 1995). Even thereafter, there is no convincing case to be made for state formation, nor even necessarily for the formation

Proceedings of the British Academy, **110**, 39–68, © The British Academy 2001.

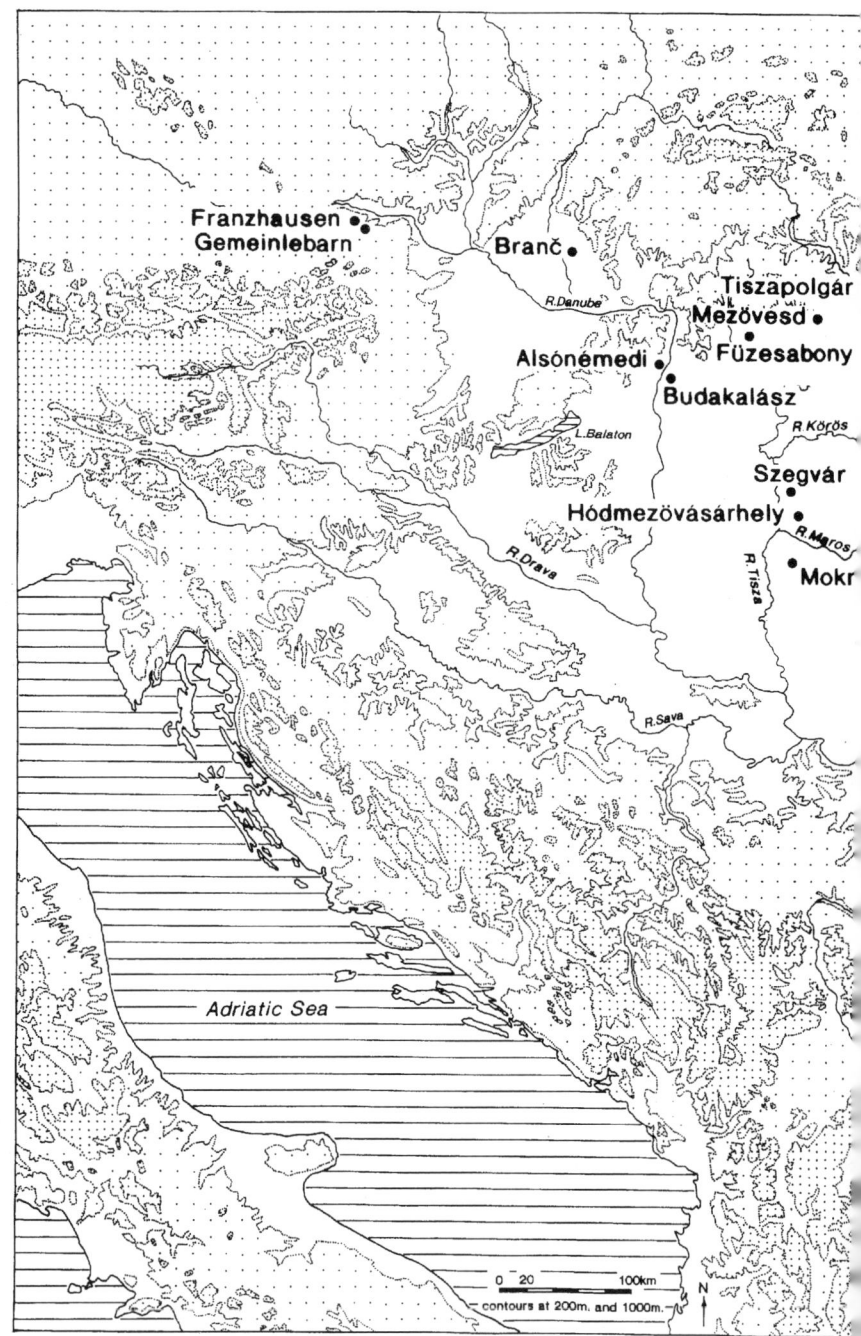

Figure 1. Map showing the principal sites mentioned in central-east Europe.

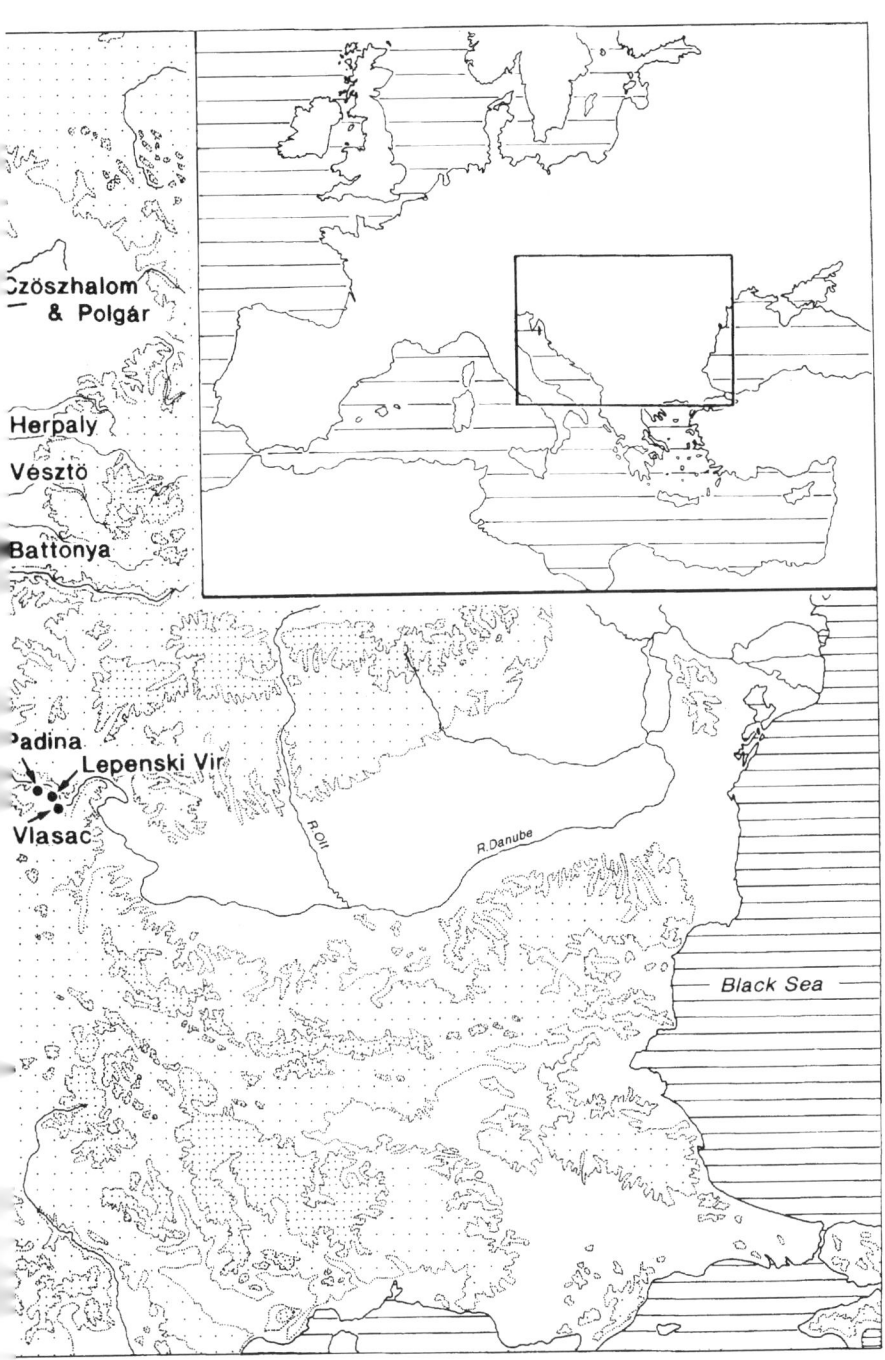

of elaborate or large-scale chiefdoms, in temperate Europe before the arrival of the Romans.

Why should this have been so? One recent discussion has explored a 'world system' approach to the relationship between temperate Europe and Mediterranean Europe and beyond from the second into the first millennium BC, to argue that the binding of temperate Europe as a periphery to the core of the Mediterranean prevented further social development (Kristiansen 1998). There can be several objections to this. It assumes that such contacts as can be discerned in the second millennium BC are already part of a core–periphery relationship, and it presumes that state formation would somehow automatically have followed, other things being equal. It also fails to explain why social formations in temperate Europe were not more elaborate by the second millennium BC in the first place. Another recent discussion is much closer to the point: 'The contrast between the Bronze Age societies of Europe and the Near East . . . is a fundamental one: they were constructed on quite different principles. European societies were actively resisting commodification, not beginning a process of convergence' (Sherratt 1994: 343).

Further explanation of the difference of the temperate European sequence has been of two kinds. One processual or modernist approach was to draw attention to the differing distributions of fertile land in Mesopotamia and temperate Europe, linked to the operation of regional exchange systems; in Mesopotamia the combination of competition for 'access to high-yielding irrigated land' with 'increasing large-scale exchange' led to social stratification by the fourth millennium BC, whereas 'the greater uniformity of the temperate European landscape deferred the emergence of such rigid forms of inequality for another 3,000 years' (Sherratt 1982: 24). The same author (Sherratt 1995: 17–19) subsequently abandoned a 'mud, reeds and people' model in favour of the more glamorous role of the Mesopotamian lowland in connecting and exploiting the exchange potential of distant areas, highland and coastal. Neither version, however, really explains very much about conditions in Europe itself. Both highly fertile land and exchangeable objects were available in Europe; with such presumed preconditions, one might, other things being equal, have expected more development on the Near Eastern model.

The other response has been to offer a series of lesser 'grand narratives', which emphasize the patterned nature of change in temperate Europe as intensifying steadily and gradually through time, though at a lesser rate than in the Near East. These have been of several different kinds, but share essentially the same reliance on notions of steady change through time. They are all problematic. They offer large-scale models, some of which may anyway be flawed in their conception, and they do not fit easily with the archaeological evidence on the ground.

One example has been modelling of stages of social 'evolution', from bands to tribes to chiefdoms (and on to states, even if not in the case of Europe). It was suggested over twenty-five years ago, for example, that chiefdoms existed in Late Neolithic southern England in the third millennium BC (Renfrew 1973), and there have been more recent suggestions for ranked chiefdoms in northern Denmark by the Early Bronze Age, following quite prolonged earlier development from simple ranking with 'Big Man' features to status rivalry without political hierarchy (Earle 1997). It remains a moot point whether it even makes sense to reduce complex and diverse phenomena to simplifying labels such as chiefdoms in the first place (Drennan and Uribe 1987), and alternatives which do not put such emphasis on social difference are often available (e.g. Whittle 1997). There is the further problem that putative chiefdoms, once supposedly arrived, did not further develop, and once again one can note the observation made some time ago that in Europe after the introduction of farming in the seventh millennium BC 'the long intervening period cannot adequately be described by a simple evolutionary succession of increasingly ranked societies' (Sherratt 1982: 14). In one of the most sophisticated examples of 'social evolutionary' modelling, developed for peninsular Italy, a Big Man society was still seen as characterizing the early second millennium BC, to be followed by a semi-stratified tribal confederacy (Robb 1994: Table 1), while it has also been suggested that small-scale chiefdoms did not appear in central Europe until the earlier second millennium BC, towards the end of the Early Bronze Age (S.J. Shennan 1993: 152).

Another example of this kind of modelling, in the sphere of economic production, is the concept of the 'secondary products revolution' (Sherratt 1981, 1987), according to which traction for the plough and wheeled vehicles, the production and consumption of milk and alcohol, and textiles were part of a later wave or waves of diffusion from the Near East, which helped to create special conditions of transformation in Europe from the fourth millennium BC onwards (Sherratt 1994). The model has been criticized by its own author as an example of 'block thinking' (Sherratt 1995: 6). Some innovations may have been present earlier in the Neolithic, including milk drinking, though the introduction of the wheel (Bakker *et al*. 1999) and perhaps the expansion of woollen textiles may be correctly described. The date for ploughing is controversial; the possibility is open that it was present much earlier, while a recent synthesis from the Alpine foreland, where the conditions of organic preservation are particularly good, suggests it could have come in rather later than predicted, gradually after 3000 BC (Jacomet and Kreuz 1999: Abb. 11.36). It remains unclear, however, what scale of effects these now more piecemeal innovations may have had in particular places. In the Alpine foreland area of south-west Germany and north-west Switzerland, for example, ploughs may be connected with a gradual shift to fixed fields or plots (Jacomet and Kreuz 1999: Abb. 11.36). Wheeled vehicles, perhaps in this instance small carts, can be documented in the Corded Ware horizon (early third millennium BC), but their specific impact on the economy is quite unclear; the overall picture of landscape development is one of very slow change through time, with clearance size and duration and the extent of cereal cultivation still on the increase in the second millennium BC (Rösch 1993).

A more recent model put emphasis on the central concepts of *domus* and *agrios*, the *domus* with its nurturing, inturned qualities being both metaphor and mechanism for social domestication, played out against the ideas of the wild inherent in the *agrios*; in time — in central and western Europe largely from the Corded Ware horizon of the third millennium BC onwards — values associated with the male world of the *agrios* came to dominate social relations (Hodder 1990). As many critics have pointed out, this binary model owes much to a simple kind of structuralism, and the model is forced on to the evidence on a continental scale, with little regard for regional variation or individual situations. A newer version has brought in the dramatic evidence of the Ice Man in relation to a discussion of the links between structure and agency (Hodder 1999: 138–44; Hodder 2000: 27). Due attention is given to the individualism of the man in the ice, but it is argued that he can be characterized above all by his independence and self-sufficiency; this allows him to be seen not only as an individual, it is argued, but as standing for the world of male-oriented independence and violence — the *agrios* — which was 'antithetical to societies based on a corporate sense of lineage and dependency and symbolised by the domestic hearth' (Hodder 1999: 144) and was becoming more important in the later fourth millennium BC in central Europe (Hodder 1999: Fig. 8.2). It is not inherently implausible that one individual could stand for a whole social formation (M. Strathern 1992, on the Pacific Garia; and see below), but interdependency is little discussed, nor the strong possibility that the Ice Man was a routine, familiar figure (Whittle 1996: 315–17; cf. Dickson 1995). Many before him had gone into and over high places. His weapons may have been far more for personal defence, including against animals, than for personal aggrandizement or aggression. The evidence for interpersonal violence in periods before the Bronze Age warrior has indeed increased in recent years (Carman & Harding 1999), but it goes at least as far back as the pre-agricultural period of the Mesolithic, and the negative evidence of a lack of injuries and an absence of specialized categories of weapons in many areas over long periods of time is also striking (Chapman 1999, 105–6); interpersonal violence of various kinds was presumably endemic in both hunter-gatherer and agricultural societies, but it did not certainly intensify through the millennia before the Bronze Age. Nor, in terms of his tools and equipment in general, was the Ice Man necessarily especially or unusually self-sufficient, as assemblages of people who lived in cold conditions in the Upper Palaeolithic seem to indicate, as at Pavilland, Sunghir, or Grimaldi (Gamble 1999).

In a less generalized fashion, but in a similar vein, it has been argued by several other authors that the nature of gender relations altered from the third to the second millennium BC. The outcome, in this view, was the appearance of a warrior ideology in the mid-second millennium BC (Treherne 1995), and beginnings have been sought in gender distinctions claimed in Corded Ware mortuary practices from the earlier third millennium BC onwards, leading to a process whereby male activities and associations were increasingly emphasized, with a concomitant downplaying of the value of women (S.J. Shennan 1993: 149), to the point where some women buried with valuable items in the Early Bronze Age of the earlier second millennium BC have been regarded as possessing such valuables by virtue of their husbands' wealth or as bridewealth in a pattern of exogamy (S.E. Shennan 1975). A related model has been argued in detail for Italy (Robb 1994). As set out in more detail below, one of the problems with this view in central Europe is the instability of representations of gender relations in mortuary ritual from place to place and through time (S.J. Shennan 1994: 124–5). In one account, the development of the use of 'secondary products' should have led to an *increase* in female power (Chapman 1997a: 137). There was much variation within the vast Corded Ware area, and cemeteries in central Europe in the second millennium BC (e.g. O'Shea 1996; Rega 1997; S.E. Shennan 1975, 1982) have much in common with earlier examples of the Early and Middle Copper Age of the later fifth and earlier fourth millennia BC (e.g. Chapman 1997a; Sofaer Derevenski 1997, 2000), of the Neolithic earlier still, back to the sixth millennium BC (Jeunesse 1997), and even indeed of the Late Mesolithic of the late seventh millennium (Radovanović 1996a, 1996b). The treatment of women in death can be distinguished from that of men, but age and life process are at least as important as biological sex (Sofaer Derevenski 1997, 2000), and the scale of difference is hardly ever extreme. It is not clear anyway that we should be seeking a simple pattern. In another context, Marilyn Strathern has urged that it may not be helpful at all to try to reduce complex situations to one dimension only, be it complementarity, dominance, or separation, because 'there is no single relationship' (M. Strathern 1987: 29). In one example, the Hua of Eastern Highland New Guinea, Meigs (1990) has emphasized a threefold male ideology, of accentuated chauvinism on the one hand, but of envy of women and of complementary interdependence on the other.

DAILY LIVES AND MORAL NETWORKS

Something remains missing from most of these narratives. It is pattern rather than texture which is being described, and explanation is generally top-down rather than bottom-up (cf. Shanks 1999: 3). We also need a much better sense of the textures and styles of past lives, and the conditions in which people understood and lived out their existence. This is not just an appeal for more detail, nor a substitution of history from the outside by history from the inside. The way in which daily lives were led, and how they were guided not only by 'forward-looking intentionality' (Hodder 2000: 23) but also by values and codes of behaviour, need to be added to the account. Looking in these ways, we may come to see different kinds of history.

Interpretive archaeology has been discussing the role of individuals since the 1980s (cf. Barrett 1994; Hodder 1986), and there is currently much debate about agency (e.g. Dobres and Robb 2000). It has been stressed that agency has been used in many overlapping ways (Dobres and Robb 2000), and is not to be equated simply with individuals (e.g. Barrett 2000: 60), but in practice many discussions of this kind have had the effect of concentrating on individuals (e.g. Hodder 1999: 133–7; 2000). This enterprise has been only partially successful, as the example of the Ice Man already discussed shows. I have set out elsewhere a short account of approaches to the individual (Whittle 1998b). In essence, with the exception of some gender-oriented narratives (e.g. Tringham 1991), what have been proposed so far are mostly disembodied agents, lacking faces, identity, motivation, or values (see also Hodder 1999: 133–7). What has arrived is a sense of individuality, in the terms of Rapport (1996; cf. Rapport 1997), rather than of individualism.

Drawing on Bourdieu (and beyond him Mauss), and other sources, there has been rather more successful, though still rather general, discussion of the *habitus*, the setting of existence in which habits and unthinking bodily action maintain a sense of understanding of the world (Gosden 1999: 124–7). An overlapping recent strand has been the 'dwelling perspective', derived in large part ultimately from Merleau-Ponty (Tilley 1994: 12–14) and Heidegger (Thomas 1996), and further considered in a series of important papers by Ingold (1993, 1995, 1996; cf. Harris 1998). This, parallel to the idea of *habitus*, discusses how people act in a world which is never separate, pre-formed, or a prior given.

The idea of individualism and the dwelling perspective begin to suggest more of the texture of lives missing from so many 'grand narrative' accounts of the various kinds already outlined above, but even these are not enough on their own. A broader approach is now required, and I will briefly sketch what I believe should be some of its constituent elements, before returning to the central European evidence. These elements include the notion of people taking a view in the world rather than of making a view of it; what goes without saying; the non-linearity of thought; the overlap of competing concepts or hybridization; the instability of individual identities; flows or networks of interaction and exchange and especially open systems; and finally shared values within a moral network.

The dwelling perspective (Ingold 1993, 1995, 1996) seeks to give a better sense of how people get on in a world which is not pre-given, 'taking the human condition to be that of a being enmeshed from the start, like other creatures, in an active, practical and perceptual engagement with constituents of the dwelt-in world' (Ingold 1996: 120–1); 'apprehending the world is not a matter of con-struction but of engagement, not of building but of dwelling, not of making a view *of* the world but of taking up a view *in* it' (Ingold 1996: 121); and finally, 'knowledge of the world is gained by moving about in it, exploring it, attending to it, ever alert to the signs by which it is revealed' (Ingold 1996: 141). This kind of approach conveys well a sense of people *attending to* their world, in ways which may not have changed much over long periods of time. As Bloch has noted (1998: 5), we are sometimes guilty of seeking too much diversity, and as already argued above, the effect of certain technological innovations may have had limited impact on the ground.

The approach, however, is incomplete. It seems to give insufficient attention to learning and to socialization. These may be long, almost unconscious or casual processes during childhood (e.g. Mead 1943), but there are also stages, such as initiation, when instruction may be much more direct, and when, as among the Hua, already cited (Meigs 1990), gender ideologies may be at their most accentuated. The approach seems to give insufficient attention to the weight of collective tradition or culture (cf. Sahlins 1999) — however that may be taken up or contested by individuals — which may affect *how* people acquire 'the skills for direct perceptual engagement' (Ingold 1996: 142) with the world. It seems an extreme claim that people never make a view of the world, and it is possible to propose that people act at different times and in different situations from varied perspectives. The dwelling perspective is best at giving a sense of the flow of life, but less satisfactory perhaps at showing how people cope with situations of change (such as the first intake of the Great Hungarian Plain, for example) or with innovations.

The dwelling perspective also evokes a rather active, conscious kind of attention to the world. This may not be how people think all the time. Bloch (1992) has explored a sense of 'what goes without saying' in Zafimaniry society in Madagascar — attitudes and beliefs, rooted in practice and material experi-ence, which are central to people but which seem so obvious that explanation of them to outsiders seems pointless. These concern ideas about such subjects as people themselves, trees, sex, gender, and houses. The relevance of elaborate house decoration is not easily put into words, as it is simply part of the right way to treat the living and growing house (Bloch 1995). Not all ideas need necessarily be seen as rooted in the practice and material experience of a single generation, since the concept of the maturing house relies on a considerable passage of time and presumably some active transmission from generation to generation. However it is passed on, this is also a powerful model for thinking about enduring beliefs over very long periods of time (even though the historical Malagasy situation has been far from static or timeless: Bloch 1998).

Another dimension of this study was a use of 'connectionism' (see also Bloch 1993), or the non-linear ways in which people often actually seem to think. Speed of thought and the ability to react instantly in different social situations suggest that thought can be held in central nodes or concepts, such as again in the Zafimaniry case to do with what people are like and how they mature, the differences and similarities between men and women, and what good marriages, trees and wood, and houses are like (Bloch 1992). In the Zafimaniry case, these could be seen to make up some sort of coherent world-view, though that might never or rarely be expressed as a single unified whole. But the compartmentalization is potentially highly significant, since it raises the possibility in other cases of non-congruence between separate nodes of thought. In another context (of discussion of post-1989 Europe), it has been proposed that this is just how we often do think, a pattern or mixture which has been called hybridization (Latour 1993). At the present time, we may be said to retain elements of pre-modern, modern and post-modern thought (Latour 1993: Fig. 5.1). This model certainly fits the discipline of archaeology in its present state and, although very general, may be a useful way to think about modes of thought in the past across horizons of change and over long periods of time.

Apart from borrowing from structuration theory (principally of Giddens), interpretive archaeology has so far developed little by way of a theory of the individual (Whittle 1998b). There is now some welcome emphasis in the more recent gender-oriented archaeological literature on life process (e.g. Sofaer Derevenski 1997: 887, 2000), whereby notions of biological sex and gender must be combined with age stages to produce a sense of identities that change through life. Rather less attention has been given to wider anthropological writing on the subject. While post-modernist archaeology has dealt with a very restricted concept of the atomized, universal individual, 'out there' notions of the individual seem to vary dramatically widely. In parts of southern India, for example, gender seems fixed and stable, based on bodily difference between men and women and their capacity for procreation, whereas in Melanesia, gender is performative, shifting, and contextually defined, depending on notions of exchange of substances or parts of persons and on the idea of the partible person (Busby 1997). In the Indian case the person may be seen as 'internally whole, but with a fluid and permeable boundary', while in the Melanesian case, the person may be 'internally divided and partible', with a mosaic of male and female substances which internally divide up the body into differently gendered parts (Busby 1997: 269–70); the body can be a microcosm of relations (M. Strathern 1988: 131), lacking stability, and extending to include objects which are or once were parts of relationships. Among the

'Are'are, the person may even be said to be made up of persons or different substances (Barraud *et al.* 1994; M. Strathern 1996: 526). In the case of the Garia, individuals themselves can stand at times for the whole society, in ways unfamiliar to a western way of thought in which the individual is normally defined in relation to a wider whole (M. Strathern 1992). Such a lack of fixity and such an ambiguity of the individual cannot be assumed as universal, as the south Indian case shows, but it is a powerful notion, and one that may be useful in thinking again about long-enduring situations in prehistoric Europe, in which individuals, in contrast to the situation in the Near East, were not tied into determined roles and identities. Something of this approach has been applied to the central European evidence, but chiefly with reference to artefacts rather than people (Chapman 1996, 2000b).

Although the Garia individual may at times be the whole, individuals have also to be set in context. They belong to networks or flows of relation, exchange, and interaction (M. Strathern 1996). These may be in part constituted by kin and closed descent groups, but there is an equally if not more important place for co-residents, friends, allies, and neighbours. It took anthropology a long time to rid itself of the notion of the dominance of the unilineal, closed descent group (Kuper 1988) and it is time for interpretive archaeology to catch up. Potentially endless networks of relations were formerly a problem for anthropologists, but only when considered in isolation (M. Strathern 1996: 529–30). Far greater attention could be given, for example, to open, bilateral systems of descent (e.g. Rivière 1995; cf. Taylor 1996: 207), which could be considered alongside the growing importance attached by interpretive archaeology to places and a sense of place in the landscape, which if not necessarily able to cut the network would at least serve to punctuate it. We will come back to both dimensions in further discussion below of the Great Hungarian Plain.

Most of the individuals sketched so far by an interpretive, post-modernist archaeology lack any or much sense of shared values. While the *domus* can indeed stand as a quite rare example of an explicitly formulated value system that includes everyone involved, the same cannot be said of the *agrios*, which only seems to have affected adult men (Hodder 1990). The Ice Man's self-sufficiency is enough to make him stand as a representative of a whole ideology (Hodder 1999). I have argued instead that there was a long-lived set of values in the European Neolithic, incorporating ideals of participation, sharing, non-accumulation, and commonality, but also the pursuit of prowess (Whittle 1996). This is not to claim that values in the European Neolithic were uniform in all times and places, nor that actual behaviour always accorded with such ideals. Nor is it to restate a Durkheimian position in which society makes its individuals. But there is socialization, and each individual does not take up a view in the world *de novo* and unaffected by others. Analogies suggest a wide range of possibilities. In an Aristotelian sense of being concerned for the response of others, it is possible to see exchange, for example, as a moral activity (Hagen 1999). In arguing that exchange among the Maneo of eastern Indonesia has in this sense a vital moral dimension, however, not reducible to its social effects, Hagen has suggested that the Maneo are not guided by specific moral principles such as would mandate sharing (1999: 362). In many cases in the Mediterranean world, individual personal responsibility is held in the 'triangle of honor, shame and luck' (Douglas and Isherwood 1996: 23). Referring to the Kabyles of Algeria, Bourdieu (1977: 48) has observed:

> The ethic of honour is the self-interest ethic of social formations, groups or classes in whose patrimony symbolic capital figures prominently. Only total unawareness of the terrible and permanent loss which a slur on the honour of the women of the lineage can represent could lead one to see obedience to an ethical or juridical rule as the principle of the actions intended to prevent, conceal or make good the outrage.

In other cases, however, imagined moral life and moral language seem to be more centrally linked to an ethical sense, tied to ideas of the person and identity, and generating powerful emotions within a shared value system. Among the Amuesha people of central Peru, greediness and meanness are regarded as immoral, irrational, and anti-social, and 'power is legitimate only when its holders are seen as loving, compassionate and generous life-givers' (Santos-Granero 1991: 229). Among the Rauto people in Melanesia, ceremonial exchange carried out in the right way creates a 'cultural landscape of memory and emotion', and the emotions generated can be considered as a kind of moral perception; emotions 'define and render compelling a particular moral stance towards life', the result of choice between this and more individualistic alternatives (Maschio 1998: 86, 97). Among the Western Apache, as described by Basso (1984), certain historical narratives served to link past events to named places, in stories which 'stalked' their listeners with their moral force; and native American appropriation of the landscape has been envisaged as by 'an act of imagination which is moral and kind' (Momaday 1976: 80). Among the Etoro of Papua New Guinea, however, male-dominated social inequality is constructed as a moral hierarchy or hierarchy of virtue, grounded in cosmology and worldview (Kelly 1993); a moral sense need not in itself be neutral.

The concept of a 'moral community' was used in a discussion of the Nuer, to connote those participating in a common value system (Johnson 1994: 327–9), as well as with reference to the Amuesha (Santos-Granero 1991: 119); the related concept of mutuality has also been discussed by Moore (1988) and Gosden (1994), and the practice of 'moral coalitions', though with a greater sense of conflict between sets of gender-based values, has been discussed by Robb (1994). Adapting these ideas, the idea of a moral network can be proposed; the extent to which others are involved and affected may define the moral network, which like other networks (Latour 1993: 117–20) is liable to remain local at all points but may also be open and unbounded. Values may be seen to act as sanctions on behaviour; there were limits to what individuals or limited interest groups could attempt or hope to get away with, and there may have been limits, within this perspective, to what they could conceive as possible.

It is not possible to discuss all these possibilities further in great detail here. In the second half of this chapter I want to concentrate on individuals in their social settings, within the perspective just outlined. In doing this, I will use burial evidence quite extensively alongside other data, and that brings of course many of its own problems of interpretation (Parker Pearson 1999). At the same time, the treatment of the dead in burial grounds shows both a very long-lived commonality and a lack of pronounced differentiation in mortuary rites. The aim is to use these two themes to show something of both the texture of daily lives and the things that endured.

WHAT INDIVIDUALS WERE LIKE:
CHANGING SETTINGS ON THE PLAIN, *c.* 6000–4000 BC

Returning to the central European evidence, specifically for the Great Hungarian Plain, I want to repeat that episodic phases of aggregation and intensification did not lead, as in the Near East, to progressively entrenched social differentiation. I want to suggest that this long-term European pattern can best be viewed through the nature of daily lives and experience, shared values in moral networks, and unstable or non-fixed individual identities. Long-term process resides in the complexity of daily lives; shared values mediated trends to aggrandizement and differentiation; and identities that were not fixed and at times ambiguous prevented people from being tied into roles and relationships in which they could be permanently dominated by others. Here are the beginnings of an explanation of the long-term patterns which I am claiming. In my first examples, I emphasize the settings and contexts of daily lives, and the shifting nature of individual identities, and in the second set of case studies I emphasize shared and enduring values as seen in mortuary practices widely separated in time.

The settlement sequence on the Great Hungarian Plain is one of the richest and in many ways best-documented such sequences in Europe as a whole. It has been intensively if episodically investigated for much of the past century, and has been described and discussed many times (in the English literature: Chapman 1994, 1997a, 1997b, 1997c, 2000a, 2000b; Sherratt 1982). The plain appears to have been largely empty at the onset of the Neolithic around 6000 BC, with indigenous populations around some of its fringes but not apparently within it (Whittle 1998a and references). Whether by colonization from the south or by processes of acculturation or enculturation involving the indigenous regional population (Whittle 1998a), the first Neolithic communities appeared in the southern part of the plain; their distribution northward was finite. This is the Körös culture of the southern part of the Great Hungarian Plain (*c*. 6000–5500 BC), part of the northern fringe of the south-east European Early Neolithic, and it represents new practices including the use of pots, domesticates, and cultivated cereals, as well as the intake of the plain itself. The settlement pattern was both riverine and dispersed. People lived off sheep and goats, cattle, other domesticates, wild game, fish, birds, and cultivated cereals. Many sites are known but it is unclear whether they all represent permanent sites or whether some are shorter-stay bases and camps; probably no one site was occupied for long and the social group at any one place of occupation may have been small, apart from during seasonal or ceremonial aggregations. There are therefore contrasts with earlier life in the Danube Gorges and in surrounding regions, but it may be an exaggeration to claim wholly new attitudes to 'nature' and 'culture' compared with what had gone before (Whittle 1998b: 473). This was a lifestyle that also involved close attention to and engagement with the environment, for stock-raising, perhaps limited cultivation on terrace edges and levees, hunting, fishing, and fowling. There was some flow of raw materials, including obsidian from the hills to the north, and so from well beyond the limits of Neolithic occupation. We do not yet understand the rhythms of this lifestyle, whether people were permanently in one place, moved with the seasons, or were captured in one place, on a seasonal basis, by flooding of the rivers, channels, and meanders whose edges they normally occupied. Occupations are marked by considerable depositions of broken pottery; one possibility is that these represent residues, many carefully placed, from seasonal or periodic aggregations of a population which was fragmented or dispersed for much of the time into much smaller units or groupings. Such places were already significant arenas.

Burials occur quite commonly on these occupations, predominantly of women and children, and there are depositions of skulls and partial remains (Chapman 1994, 2000b). Figurines present, on a literal reading, rather bland if not anonymous and mask-like faces (some could even be taken to represent people wearing masks); many are of the human female form, but the common type with elongated neck may also have phallus-like qualities (Whittle 1998b: 473–5, Abb. 1). Representations of the human form on pottery also lack faces, and emphasize stylized gesture or activity. It is possible (starting again from a literal reading) that identity was at this time ambiguous or in parts contradictory (cf. Taylor 1996), with frequent metaphorical or literal masking of the inner self, some blurring of the differences between women and men combined with different treatments in death (men's bodies presumably being frequently deposited in some way out in the landscape), and attachment to and frequent commemoration of chosen places, but also a fluidity of allegiance which allowed much coming and going in the landscape as people attended to a spectrum of social relationships and other activities.

From around 5500 BC, there was further expansion across the plain, taking Linear Pottery (AVK: *c.* 5500–5000 BC) settlement to the northern fringes of the plain. In many ways the settlement pattern continued as before, with the population dispersed through the riverine system. There was continued inter-action with people around the fringes of the plain. Pollen analysis in the north-ern hills suggests woodland clearance or interference there (Gardner 1999a, 1999b), and some of this activity close to the plain could have been carried out by people from the plain. Large-scale motorway excavations near the north edge of the plain at Füzesabony-Gubakút have recently shown the existence of longhouses, formerly not recognized until the late AVK Szakálhát phase (mainly known further south); there were also graves near the houses, notably of children as well as mature adults (Domboróczki 1997). It remains to be seen whether these structures were typical of the whole plain, and to what extent they were the centre of a sedentary existence. Structures at the early AVK site of Mezőkövesd-Mocsolyás (Kalicz & Koós 1997) lacked postholes; they were accompanied by 25 graves. One possibility is that such structures will prove to be markers of the establishment of local, short-lived sedentism.

The human face is represented on figurines, pot attachments, and pot deco-rations. Pot or clay vessel attachments from Füzesabony-Gubakút show trian-gular faces, still rather expressionless and mask-like, or even representing masks. So far, the majority of such representations come from the northerly parts of the plain (Kalicz and Makkay 1977: Abb. 4). We do not know whether such representations are of spirits, ancestor figures, or even particular individ-uals, and the nature of individualism can hardly be easily read off from them. It could be that they are something to do with the assertion of identity in the newly colonized parts of the north of the plain, and with the creation of longhouse-based relationships. The connection between house and face repre-sentation is also seen in the Szakálhát phase, but to the south. A well-known series of face pots continues the emphasis on human form, with sharper delin-eation of eyes, nose, and mouth, and schematic delineation of hair and ears. The recurrent M-motif below the face has been suggested as female, but this is open to question. The remains of at least 22 such pots were found in a burnt house at Battonya (Goldman 1978), suggesting a close relationship between the emergence of households, however constituted, and a continued interest in the definition of identity. The dead were themselves treated simply in mortuary rites, while the rupture of death may have had more dramatic effects on house-hold life, bringing it on occasion to an abrupt end by deliberate house destruc-tion (Stevanović 1997).

Just as AVK-longhouses could be seen as an enhancement of Körös structures, so the tell settlements of the Tisza culture period (*c.* 5000–4500 BC) can be considered as a development of what had gone before, leading to the local establishment of phases of sedentism and aggregation. The first tells emerged, principally in northern and eastern parts of the plain, alongside the continued use of 'open' or 'flat' sites (Raczky *et al.* 1994: colour table III). The situation seems very complex. Tells themselves varied in character, from some of the lower and broader examples further south, in which there were discernible shifts of focus through time, to the small conical tells of the eastern Herpály group, to the northerly ditched roundel of Csőszhalom, tell-like and with some occupation on the mound, but accompanied by a large open settlement a little distance away (Raczky 1987; Raczky *et al.* 1994, 1997a). Some of these sites may have been deliberately planned 'timemarks' which drew on an ideology of deep ancestral time, the aim of a deliberate strategy carried out by limited local interest-groups (Chapman 1997b, 1997c), rather than part of a grand plan to manage plains–fringes relationships including flows of goods and cattle (Sherratt 1982; cf. Shanks and Tilley 1987: 37–41). However, it is also possible that the mounds in question were rather more the outcome of social practices than a predetermined effect, the unplanned result of prolonged occupation of one place. In any case, they represent concentrations of well-constructed buildings, which seem at least in part to be connected with permanent occupation. Cattle became the most important animal. There were flows of exchange within the plain and with the fringing highland, involving perhaps both animals and material valuables (Sherratt 1982). The tells speak for continuity, aggregation, and perhaps an importance given to group ancestry, but these were not in themselves new in this period. Nor are tells found uniformly across the plain, as noted already, and individual site histories vary considerably in the intensity and episodicity of occupation; it is not yet certain that all tells were inhabited all year by the same people. Difference is thus more apparent between regions and between sites, than within sites.

Houses seem not to be markedly differentiated by size or contents (the latter now including rich assemblages of decorated pottery and figurines). The dead were again present among the living. Child burials and aurochs skulls under house walls and floors at Berettyóújfalu-Herpály (Kalicz and Raczky 1987) bound fertility and the power of wild creatures into the foundation of the house. Groups of burials, probably mostly in open spaces between houses, are known, some now in coffins; adults are well represented, including men, but children were in a clear majority at Berettyóújfalu-Herpály. Some adults were presumably therefore still disposed of out in the landscape. Grave goods are generally simple, as before, including beads, other simple ornaments, and simple tools at Vésztő-Mágor (Hegedűs and Makkay 1987), and ornaments in stone, bone, spondylus, and copper at Berettyóújfalu-Herpály (Kalicz and Raczky 1987). In the settlement near to Csőszhalom, men and women were distinguished by being lain on opposing sides of the body; women had strings of beads in marble and spondylus, on their heads and around their waists, as well as spondylus bracelets, while men had small stone axes and pairs of modified wild boar tusks. Anthropomorphic vessels, vessel attachments, and figurines were part of the rich material culture. Their forms were now very varied, from the striking sitting figurines (male as well as female) of sites such as Hódmezővásárhely-Kökénydomb, Vésztő-Mágor, and Szegvár-Tűzköves (Razcky 1987), some of which are really vessels as well as figurines, to the much more modest small figurines of a site such as Berettyóújfalu-Herpály (Kalicz and Raczky 1987: 206–8). It remains dangerous to try to read off identities from these representations, but their very diversity may speak in a general way for a greater sense of individualism.

There was thus much continuity in this phase as well as slow change. Tells did not emerge overnight, and are barely characteristic of the plain as a whole. Some of their occupants may not have been present on them all the time. They may have attended to their landscapes in very similar ways to their predecessors, though with cattle now becoming the animal of dominant concern. Individuals continued to belong to groups which had a concern for descent and belonging. But things were hardly exactly as before. Individual places were picked out for special occupation (cf. Chapman 1997c), and the business of living in closely spaced houses, even if only for seasons or for short runs of years, must have affected individuals in new ways; the effect, however, at least in specific places and at specific times, may have been to constrain individual action, at a time when figurines and burials might suggest greater individualism.

Around the middle of the fifth millennium BC the pattern alters again, and many of those sites with previously prolonged reoccupations were now either abandoned or much less intensively used. From the Early Copper Age Tiszapolgár culture through to the Late Copper Age Baden culture, there seems to have been a long phase again of dispersal of small sites across the landscape. In the Tiszapolgár culture phase (c. 4500–4000 BC), just at the point when certain sites or groups could have created or reinforced pre-eminence, few tells remained in occupation. The break was not immediate. There were proto-Tiszapolgár burials at Hódmezővásárhely-Gorzsa (Horváth 1987) and Tiszapolgár burials at Vésztő-Mágor. The population seems mainly to have been dispersed once more in scattered small units, in a lifestyle in many ways of now considerable antiquity. To our eyes at least, the most notable places in the Early and Middle Copper Age (or Bodrogkeresztúr culture, 4000–3500 BC) were burial grounds, set apart from settlements or occupations; the classic but probably atypical example is Tiszapolgár-Basatanya itself (Bognár-Kutzián 1963, 1972). Once again, this shift had been prefigured in the preceding phase; some of the burials at Berettyóújfalu-Herpály, for example (Kalicz and Raczky 1987), had been placed at a distance from the tell. While it has been suggested

Alasdair Whittle

that competing limited interest-groups would actively have sought new ways of promoting themselves, and would thus have sought to make displays in this period through mortuary rites separate from occupations (Chapman 1997b), other explanations may apply. The costs of maintaining tell existence may have become too high (cf. Bogucki 1996), in social as well as economic terms. In another context, among the Foi of Papua New Guinea, considerable tension from the demands of close communal living has been recorded, to be contrasted with the greater freedom of more independent existence (Weiner 1991: 78). The flow of materials and goods through tells and open sites in this period (cf. Sherratt 1982: Fig. 2.5) may have produced tensions at odds with an otherwise communal ethos (cf. Hagen 1999), to be resolved once more by fission among the living. There are burial grounds of varying sizes, the smaller ones perhaps serving more local populations than the largest example, Tiszapolgár-Basatanya itself (Bognár-Kutzián 1963); the northerly position of that site in relation to flows of copper into the plain may have been significant (Sherratt 1982). While some sites therefore may have been the location of a greater range of artefacts, and been longer-lived, there is little other sign of differentiation within the length and breadth of the plain (Bognár-Kutzián 1972). Burial grounds both large and small seem to promote an ideology of commonality, now prolonged among the close community of the dead after its practice among the aggregations of the living on tells and large open sites in the preceding Tisza phase.

Not all Tiszapolgár mortuary rites were identical. Nor was the situation necessarily static. In the Early Copper Age some of the richest sites may have been near or on the edge of the plain, well placed for exploiting movements of copper (Sherratt 1982). The small group excavated at the Vésztő-Mágor tell to the south of the Körös River were more traditionally furnished than elsewhere, with less variation in grave goods, but still some gender differentiation (Chapman 1997a: 143). From other instances, however, there seems to have been a very widely distributed common way of doing things. A proto-Tiszapolgár female grave at Hódmezővásárhely-Gorzsa, east of the Tisza and north of the Maros (Horváth 1987: Fig. 23), was provided with a pot behind the head, abundant beads probably formerly in strings on the legs, hips, and neck, and a bracelet on the right arm. By the early Tiszapolgár phase at Polgár-Nagy Kasziba in the north of the plain (Raczky *et al.* 1997b), there were a male, a female, and two child graves (probably part of a larger burial ground) with differentiation by body side and by some of the grave goods, in ways familiar from Tiszapolgár-Basatanya, while both the woman and the man were accompanied by many pots.

Tiszapolgár-Basatanya itself has been the most analysed, since it is well published (including Chapman 1997a; Meisenheimer 1989; Sherratt 1982; Sofaer Derevenski 1997, 2000). Its use ran from the Tiszapolgár phase on into the succeeding Bodrogkeresztúr phase, on into the fourth millennium BC; there were over 150 graves in all, mainly individual inhumations. Women, men, and children are represented. The dead were set in rows and their graves were probably individually marked, their positions thereby subsequently respected over long periods of time. Though there is a wide range of grave goods (Table 1), there is overall no clear sign of major material differentiation between individuals or groupings (the latter explored by Meisenheimer 1989) within the burial ground. There was clearly some emphasis put on gender differences, as emphasized by recent analyses (Chapman 1997a; Sofaer Derevenski 1997, 2000). These change between the Tiszapolgár and Bodrogkeresztúr phases, though it may be rather artificial to contrast only two such large blocks of time without trying to take account of changes from generation to generation (cf. Meisenheimer 1989). In Sofaer Derevenski's accounts (1997: 887, 2000), the various objects in question are linked in the Tiszapolgár phase not only to gender but to age, creating a strong sense of life process, whereas in the Bodrogkeresztúr phase there is a greater emphasis simply on female:male difference (reinforced by the slightly different analysis of Chapman 1997a: 138–43). While there is thus difference, there is hardly, in crude terms, discernible inequality evident in the mortuary rites, since women's graves can be as abundantly furnished with goods as men's, and children (at least in the Tiszapolgár phase when they are more common) seem often to be treated in anticipation of their future development. There seems to be increased emphasis throughout on the dead as though they were living: a performative view of people. Thus women, men, and children all receive food remains and quite abundant and varied pottery, though relative numbers may vary depending on age and gender (Sofaer Derevenski 1997: Fig. 2), and there are tools and other objects to do with dress and appearance. The last sight of the dead in the grave

Table 1. The range of grave goods and their principal gender associations in the Early Copper Age phase (Tiszapolgár culture) at the burial ground of Tiszapolgár-Basatanya, Hungary (modified after Chapman 1997a: Table 10.3). For further details of age categories, see Sofaer Derevenski (2000).

Social category	Grave goods
Child only	Flint scraper, obsidian
Adult male only	Wild animal bones, snails, complete dog, limestone disc, loom-weight, ochre lumps
Adult female only	Deer tooth, bone spoon, pebble
Adult male and child only	Shed antler, boar's tusk, cattle metatarsal, pig mandible, domestic animal bones, copper bracelet
Adult female and child only	Aurochs bone
Adult male and adult female	Mussels, cattle bones, bone awl
All categories	Pottery, antler artefacts, flint blade, ground stone, limestone beads, copper bead and ring, fire

was of individuals furnished, in ways appropriate to their gender and age, for full, active participation in social life and the maintenance of long-established commonality.

WHAT THE DEAD WERE LIKE:
FURTHER EARLY AND LATE EXAMPLES

I have concentrated so far on the shifting sense of individualism in the changing settings in which daily existence was lived out. I want now to use two further examples, from the beginning and end of this long sequence, to suggest from a study of mortuary rites that many shared values seem to have endured for very long periods of time; in Barrett's terms (2000: 65), while structural conditions may have changed quite frequently, structuring principles may have altered rather less. These examples can be combined with others already considered above.

The nature of late hunter-gatherer communities in the Danube Gorges is a complex issue and this brief account is not intended to offer mere simplification, but the example provides a useful and important starting-point. Even the chronology of development is uncertain; I follow Radovanović (1996a) in seeing the most intense practices as later rather than earlier, probably at the transition from the seventh to the sixth millennium BC, and perhaps in reaction to the appearance elsewhere in the wider region of early Neolithic communities (Whittle 1996). While some communities in the gorges may have been semi-sedentary, others may have followed a more mobile lifestyle. In either case, people were engaged in their environment with subsisting, including fishing in the river and hunting in the hills. There were flows of raw materials into the gorges from north and south (Chapman 1993). People chose special places such as Lepenski Vir, Vlasac, and Padina (Radovanović 1996b), in the innermost gorges, to mark their identity and their relationship to the river and its surrounding rocks and hills, with structures, perhaps shrines, and with burials, offerings, and other depositions. By Lepenski Vir II, in the conventional chronology, the dead were placed parallel to the river and facing downstream (Radovanović 1997). By the same phase sculpted or carved boulders in structures or shrines, which had first been more abstract, had assumed ambiguous faces, between human and fish. These have plausibly been seen as based on the great sturgeon of the river (Radovanović 1997; Srejović 1972) and beyond that as ancestor figures, whose annual return upstream was awaited by the dead and presumably also by the living (Radovanović 1997). In the latter stages of Lepenski Vir, it seems that large sturgeon were not eaten. It has been tempting to present this as a unified system, revealing hunter-gatherers in tune with 'nature', as opposed to the orientations of early farmers to 'culture' and human identity (e.g. Whittle 1998a). This is too simple. There may be hybridization here already, plausibly enough in a context of regional change. There may have been multiple identitities, and more than one kind of ancestry, of creatures as well as humans, though bound together by continued attention to chosen special places.

The dead were collected in these special places, perhaps in greater numbers than elsewhere (Radovanović 1996b: Table 1). The dead could be divisible, especially adults, separate remains of skulls and jaws being notable features of Lepenski Vir I (Radovanović 1996b: Table 13). Some of the dead were provided with accompanying goods, with already some differentiation, though not marked, by gender and age (Radovanović 1996b: Tables 7–14 (see Table 2)).

Changes continued to occur through time, and something of these has been seen in the previous section. There is not the space here to complete a detailed account through the fourth and third millennia BC. Thus the Baden horizon starting in the mid-fourth millennium BC could have seen the shift or part of the shift to Indo-European language, and the arrival of wheeled vehicles (Sherratt 1981), though 'secondary products' do not seem to have affected the situation on the ground to any great extent in terms of possible intensification or complexification. There were some dramatically new mortuary rites, including the burials of people with cattle pairs at Alsónémedi, off the plain near the Danube by Budapest (summarized in Whittle 1996: 122–6). From the perspective argued here, however, we could just as well stress the now very familiar kinds of mortuary rites seen in the rest of the burial ground at Alsónémedi. A comparable situation can be seen in the largest burial ground of this period, at Budakalász (Whittle 1996: 124 and references). A more recent characterization has talked of 'kaleidoscopic potential for fine-tuned and diverse statements about the living and the dead' and 'subtly varied identities' in the deployment of grave goods in this cemetery (Chapman 2000b:164 and Fig. 5.11). Without therefore ignoring this and other intervening horizons of change (see Chapman 1997a, 2000b; S.J. Shennan 1993), the final examples here will come from the early second millennium BC. There has been a tendency for Neolithic specialists to end their interpretations in the fourth or third millennium, and for Bronze Age specialists to begin in the late third millennium. It is instructive to take a longer view.

By this date, the late third running into the earlier second millennium BC, there had been further changes including the development of abundant bronze production and enclosed upland sites, the latter not necessarily to be regarded as fortifications (Harding 2000: 294); tells reoccupied or created on the Great Hungarian Plain do not appear in the main to be defended or placed with regard to defence or remoteness. In this horizon, the tradition of cemetery burial was not only maintained on the Great Hungarian Plain, but found widely over much of the area between the Rhine and the Vistula (Harding 2000:

Table 2. Selected grave goods and their principal age and gender associations in late levels at Vlasac, and in Lepenski Vir I (later levels) and II, in the Danube Gorges.

Artefact	Vlasac	LV I	LV II
Bone awl	–	f sen	f mat
Bone projectile	f mat	–	–
Bone tool	f mat	–	f mat
Antler tool	m mat	–	–
	f adu	–	–
	f mat	–	–
Boar's tusk tool	f mat	–	–
Flint	f mat	m juv	–
Decorated stone	m adu	–	–
Decorated bone	f mat	–	–
Necklace	inf	–	f mat
			f adu
Cyprinidae teeth	m mat	–	–
	m adu	–	–
	f mat	–	–
	f adu	–	–
	inf		

Note: f: female; m: male; inf: infant; juv: juvenile; adu: adult; mat: mature; sen: senior.
Source: After Radovanović 1996: Table 8.

76). While barrow burial in some peripheral areas further north may suggest greater extremes of wealth, sites such as Helmsdorf and Leubingen in Thüringen being often-cited examples, cemeteries of mainly individual inhumations seem to project a different kind of ideology, though this need not have been stable or uniform across such a wide area. Allegiance was given, at least in death, in part to a collective ideal, and burial grounds can include literally hundreds of graves, for example at Franzhausen in the Danube Valley west of Vienna (Neugebauer & Neugebauer 1997). Attention was also given to other groupings, perhaps on the basis of kinship or residence, since many burial grounds appear to have internal areas in simultaneous use; Gemeinlebarn, Singen, and Mokrin are examples (Harding 2000, with references). Women, men, and children continued to be represented, individual graves being presumably marked, and successive burials often being made in rows to respect previous interments. There was normally distinction made between men and women in terms of body side.

Much analysis has been carried out of variations in the provision of grave goods (e.g. Bátora 1991; Kadrow 1994; O'Shea 1996; Rega 1997; S.E. Shennan 1975, 1982; cf. Harding 2000: 394–8). While there are graves with no goods, as in a sizeable percentage at Iwanowice near Kraków (Kadrow 1994), or in some of the graves at Mokrin (Rega 1997: 231), many contain something, and as in earlier times pots were recurrent. In many accounts, the amounts or kinds of goods deposited with the dead have been the dominant consideration. The wider context sketched above, and the circumstances in which different kinds of deposition might have been appropriate (Parker Pearson 1999), are both often neglected. There is indeed variation, and it is possible that the range of variation was greater than in earlier times. At Brančʼ, Slovakia, there were more richly endowed graves of women than of men, and the explanations discussed so far have included a display of the wealth of husbands, descent through the female line, bridewealth, and polygyny (S.E. Shennan 1975: 286). In southern Hungary and northern Serbia, it is proposed that whereas the treatment and disposition of the body were strongly normative, more variation relating to status can be seen in grave good variation (O'Shea 1996: 187). Such a combination might rather be considered problematic, though it could reflect various kinds of tension, not inconsistent with the discussion above of overlapping points of view about the world. Some of the variation is ascribed to life process, to changing roles through successive changes in life (O'Shea 1996: 276–83; cf. Rega 1997), which is familiar from the situation in the Early Copper Age at Tiszapolgár-Basatanya, but some, such as sashes for women and head ornaments and weapons for men, to hereditary and other vertical social differentiation (O'Shea 1996: 283–94) (see Table 3).

There is little need to challenge the demonstration of variability in the provision of grave goods, but the total context does cast doubt on the scale of differentiation and the rather mechanical scoring and separation of individual elements of the totality of mortuary treatment. It is also telling that analysis of the Maros area ends with discussion of 'leveling rules' that maintained a 'basic egalitarian ethos' through the 800-year existence of the tells or villages in question, 'enforced by the power of obligation, reciprocity and kinship' (O'Shea 1996: 348–9). Nor is there any need to insist upon uniformity, but the observations that at Brančʼ, in south-west Slovakia, 'the distinctions between "rich" individuals and others are by no means absolute' (S.E. Shennan 1975: 287), and that in the early second millennium BC the detected shifts, varying

Table 3. Suggested classification of funerary distinctions at Mokrin.

Feature	Sex	Age	Social category	Type of differentiation
Cemetery burial	all	all	community membership	normative
Flexed position	all	all	community membership	normative
East-facing	all	all	community membership	normative
Weapon	m	adu	hereditary social office	vertical
Head ornament	m	adu	hereditary social office	vertical
Beaded sash	f	adu	hereditary social office	vertical
Bone needle	f	?	hereditary social office	vertical
Head ornament	f	adu	associative social position	vertical

Note: f: female; m: male; adu: adult.
Source: Selected and simplified from O'Shea 1996: Table 8.5.

from site to site with more marked examples in Mokrin and other burial grounds of the Maros area, were from minimal to moderate ranking (S.E. Shennan 1982), have particular significance in the context of this discussion of the nature of change over very long spans of time.

DIFFERENT KINDS OF HISTORY

This returns us to what I have referred to above as the moral network. Probably never quite the same from place to place and time to time, and constantly open to re-interpretation, manipulation, and indeed subversion, because we cannot reduce social interactions to single relationships, none the less shared values and ideals seem from the evidence presented to have been extremely long lived, and capable of enduring through changing conditions of existence. By the Early Bronze Age, some individuals may have had a more defined social role, though that was hardly a uniform or stable development, and older concerns for the collective of the shared burial ground, distinctions between genders, allegiance to groupings within the wider whole, changing roles through life, and participation in social interaction, especially through the medium of food and drink, remain prominent. It has been a standard assumption to suppose that power and competition for it were the major focus of people's lives, that such concerns intensify through time, and that the principal actors involved were members of limited interest-groups (e.g. Chapman 1997b: 147; cf. Bailey 2000). This seems to me to pick out a single dimension of past social interaction for undue attention, and to risk being drawn into seeking patterns of steadily intensifying change where none may exist (cf. Chapman 2000b: 167). It avoids the complexity and messiness of past lives, and the rhythms and timescales of individual existence in specific places. Tension between people may have existed throughout the central European case study explored here, to varying degrees. In the case of the Pangia area of the southern highlands of Papua New Guinea, it is argued that principles of hierarchy and unity coexisted, with 'debate, confusion and improvisation' and continuing 'experiential flux' (A. Strathern and Stewart 2000). The European evidence discussed seems to suggest a less extreme situation, but the argument here can allow a measure of what has been called 'alternating disequilibrium' (A. Strathern 1971: 11) from phase to phase or within particular phases. Overall, however, this did not lead to the establishment of a pattern of marked hierarchy, even by the earlier second millennium BC, in accordance with the many older characterizations of the sequence noted at the start of the chapter.

What I have tried to show is that giving attention to the conditions and textures of daily existence may help better to explain this long duration. This was not a history in which nothing happened and in which nothing changed; some kinds of change were recurrent, but these came at irregular intervals. At varying times there were, in some senses, cycle, reversal, and stasis. In addition, the timescale of individual lives allied to shifting senses of what it was to be an individual conditioned a state of flux within social networks. People acted in varying ways, often from habit and without thinking, and often in non-linear and overlapping or contradictory ways, in skilled and continuous engagement with their social worlds, regularly in accordance with the expectations of others, and sometimes more consciously to create new possibilities. They were also constrained by what was normally sanctioned as acceptable. This was learnt through childhood, reinforced at initiation, and encountered daily in the social taskscape, which changed through life with age and role. I argue that it was these broadly shared values, meshed with the ways in which people routinely acted out their daily lives and thought about their worlds, which enabled the long continuities of the different kind of history proposed here. Perhaps the kind of developments seen in the Near East remained for many European moral networks simply unimaginable.

Note. I am very grateful to Douglass Bailey, John Evans, Kathy Fewster, Joshua Pollard, and the organizer and editor Garry Runciman for their criticism of earlier drafts of this paper, to the conference participants for discussion and comment, and to Anthony Harding for information.

REFERENCES

BAILEY, D. 2000: *Balkan Prehistory: Exclusion, Incorporation and Identity* (London, Routledge).
BAKKER, J.A., KRUK, J., LANTING, A.E. & MILISAUSKAS, S. 1999: The earliest evidence of wheeled vehicles in Europe and the Near East. *Antiquity* 73, 778–90.
BARRAUD, C., DE COPPET, D., ITEANU, A. & JAMOUS, R. 1994: *Four Societies Viewed from the Angle of their Exchanges* (Oxford, Berg).
BARRETT, J.C. 1994: *Fragments from Antiquity: an Archaeology of Social Life in Britain, 2900–1200 BC* (Oxford, Blackwell).
BARRETT, J.C. 2000: A thesis on agency. In Dobres, M.-A. & Robb, J. (eds.), *Agency in Archaeology* (London, Routledge), 61–8.
BASSO, K.H. 1984: 'Stalking with stories': names, places, and moral narratives among the Western Apache. In Bruner, E.M. (ed.), *Text, Play and the Story: the Reconstruction of Self and Society* (Washington, DC, American Ethnological Society), 19–55.
BÁTORA, J. 1991: The reflection of economy and social structure in the cemeteries of the Chłopice-Veselé and Nitra culture. *Slovenská Archeológia* 39, 91–142.
BLOCH, M. 1992: What goes without saying: the conceptualization of Zafimaniry society. In Kuper, A. (ed.), *Conceptualizing Society* (London, Routledge), 127–46.
BLOCH, M. 1993: Domain specificity, living kinds and symbolism. In Boyer, P. (ed.), *Cognitive Aspects of Religious Symbolism* (Cambridge, Cambridge University Press), 111–19.
BLOCH, M. 1995: Questions not to ask of Malagasy carvings. In Hodder, I., Shanks, M., Alexandri, A., Buchli, V., Carmen, J., Last, J. and Lucas, G. (eds.), *Interpreting Archaeology: Finding Meaning in the Past* (London, Routledge), 212–15.

BLOCH, M.E.F. 1998: *How We Think They Think: Anthropological Approaches to Cognition, Memory, and Literacy* (Boulder, CO, Westview Press).

BOGNÁR-KUTZIÁN, I. 1963: *The Copper Age Cemetery of Tiszapolgár-Basatanya* (Budapest, Akadémiai Kiadó).

BOGNÁR-KUTZIÁN, I. 1972: *The Early Copper Age Tiszapolgár Culture in the Carpathian Basin* (Budapest, Akadémiai Kiadó).

BOGUCKI, P. 1996: Sustainable and unsustainable adaptations by early farming communities of northern Poland. *Journal of Anthropological Archaeology* 15, 289–311.

BOURDIEU, P. 1977: *Outline of a Theory of Practice* (Cambridge, Cambridge University Press).

BUSBY, C. 1997: Permeable and partible persons: a comparative analysis of gender and body in south India and Melanesia. *Journal of the Royal Anthropological Institute* 3, 261–78.

CARMAN, J. & HARDING, A. (eds.) 1999: *Ancient Warfare: Archaeological Perspectives* (Stroud, Sutton).

CHAPMAN, J. 1993: Social power in the Iron Gates Mesolithic. In Chapman, J.C. &. Dolukhanov, P. (eds.), *Cultural Transformations and Interactions in Eastern Europe* (Aldershot, Avebury), 61–106.

CHAPMAN, J. 1994: The living, the dead and the ancestors: time, life cycles and the mortuary domain in later European prehistory. In Davies, J. (ed.), *Ritual and Remembrance: Responses to Death in Human Societies* (Sheffield, Sheffield Academic Press), 40–85.

CHAPMAN, J. 1996: Enchainment, commodification and gender in the Balkan Copper Age. *Journal of European Archaeology* 4, 203–42.

CHAPMAN, J. 1997a: Changing gender relations in Hungarian prehistory. In Moore, J. & Scott, E. (eds.), *Invisible People and Processes: Writing Gender and Childhood into European Archaeology* (London, Leicester University Press), 131–49.

CHAPMAN, J. 1997b: The origin of tells in eastern Hungary. In Topping, P. (ed.), *Neolithic Landscapes* (Oxford, Oxbow), 139–64.

CHAPMAN, J. 1997c: Places as timemarks: the social construction of prehistoric landscapes in eastern Hungary. In Chapman, J. & Dolukhanov, P. (eds.), *Landscapes in Flux: Central and Eastern Europe in Antiquity* (Oxford, Oxbow), 209–30.

CHAPMAN, J. 1999: The origins of warfare in the prehistory of central and eastern Europe. In Carman, J. & Harding, A. (eds.), *Ancient Warfare: Archaeological Perspectives* (Stroud, Sutton), 101–42.

CHAPMAN, J. 2000a: Tension at funerals: social practices and the subversion of community structure in later Hungarian prehistory. In Dobres, M.-A. & Robb, J. (eds.), *Agency in Archaeology* (London, Routledge), 169–95.

CHAPMAN, J. 2000b: *Fragmentation in Archaeology: People, Places and Broken Objects in the Prehistory of South-Eastern Europe* (London, Routledge).

DICKSON, J.H. 1995: Plants and the Iceman: Ötzi's last journey. *www.gla.ac.uk/Acad/IBLS/DEEB/jd/otzi.htm.*

DOBRES, M.-A. & ROBB, J. 2000: Agency in archaeology: paradigm or platitude? In Dobres, M.-A., & Robb, J. (eds.), *Agency in Archaeology* (London, Routledge), 3–17.

DOMBORÓCZKI, L. 1997: Füzesabony-Gubakút. In Raczky, P., Kovács, T. & Anders, A. (eds.), *Utak a múltba: az M3-as autópálya régészeti leletmentései* (Budapest, Magyar Nemzeti Múzeum and Eötvös Loránd Tudományegyetem Régészettudományi Intézet), 19–27.

DOUGLAS, M. & ISHERWOOD, B. 1996: *The World of Goods: Towards an Anthropology of Consumption* (London, Routledge).

DRENNAN, R.D. & URIBE, C.A. (eds.) 1987: *Chiefdoms in the Americas* (Lanham, MD, University Press of America).

EARLE, T. 1997: *How Chiefs Come to Power: the Political Economy in Prehistory* (Stanford, Stanford University Press).

GAMBLE, C. 1999: *The Palaeolithic Societies of Europe* (Cambridge, Cambridge University Press).

GARDNER, A. 1999a: 'The Impact of Neolithic Agriculture on the Environments of South-East Europe.' PhD thesis, Cambridge University.

GARDNER, A. 1999b: The ecology of Neolithic environmental impacts — re-evaluation of existing theory using case studies from Hungary and Slovenia. *Documenta Praehistorica* 26, 163–83.

GOLDMAN, G. 1978: Gesichtsgefässe und andere Menschendarstellungen aus Battonya. *A Békés Megyei Múzeum Közleményei* 5, 13–60.

GOSDEN, C. 1994: *Social Being and Time* (Oxford, Blackwell).

GOSDEN, C. 1999: *Archaeology and Anthropology* (London, Routledge).

HAGEN, J.M. 1999: The good behind the gift: morality and exchange among the Maneo of eastern Indonesia. *Journal of the Royal Anthropological Institute* 5, 361–76.

HARDING, A. 2000: *European Societies in the Bronze Age* (Cambridge, Cambridge University Press).

HARRIS, M. 1998: The rhythm of life on the Amazon floodplain: seasonality and sociality in a riverine village. *Journal of the Royal Anthropological Institute* 4, 65–82.

HEGEDŰS, K. & MAKKAY, J. 1987: Vésztő-Mágor: a settlement of the Tisza culture. In Raczky, P. (ed.), *The Late Neolithic of the Tisza Region* (Budapest and Szolnok, Szolnok Museum), 93–111.

HODDER, I. 1986: *Reading the Past* (Cambridge, Cambridge University Press).

HODDER, I. 1990: *The Domestication of Europe* (Oxford, Blackwell).

HODDER, I. 1999: *The Archaeological Process: an Introduction* (Oxford, Blackwell).

HODDER, I. 2000: Agency and individuals in long-term process. In Dobres, M.-A. & Robb, J. (eds.), *Agency in Archaeology* (London, Routledge), 21–33.

HORVÁTH, F. 1987: Hódmező vásárhely-Gorzsa: a settlement of the Tisza culture. In Raczky, P. (ed.), *The Late Neolithic of the Tisza Region* (Budapest and Szolnok, Szolnok Museum), 33–49.

INGOLD, T. 1993: The temporality of the landscape. *World Archaeology* 25, 152–73.

INGOLD, T. 1995: Building, living, dwelling: how animals and people make themselves at home in the world. In Strathern, M. (ed.), *Shifting Contexts: Transformations in Anthropological Knowledge* (London, Routledge), 57–80.

INGOLD, T. 1996: Hunting and gathering as ways of perceiving the environment. In Ellen, R. & Fukui, K. (eds.), *Redefining Nature* (Oxford, Berg), 117–55.

JACOMET, S. & KREUZ, A. 1999: *Archäobotanik: Aufgaben, Methoden und Ergebnisse vegetations- und agrargeschichtlicher Forschung* (Stuttgart, Ulmer).

JEUNESSE, C. 1997: *Pratiques funéraires au Néolithique ancien: sépultures et nécropoles des sociétés danubiennes (5500–4900 av. J.-C.)* (Paris, Editions Errance).

JOHNSON, D.H. 1994: *Nuer Prophets: a History of Prophecy from the Upper Nile in the Nineteenth and Twentieth Centuries* (Oxford, Clarendon Press).

KADROW, S. 1994: Social structures and social evolution among Early Bronze Age communities in south-eastern Poland. *Journal of European Archaeology* 2, 229–48.

KALICZ, N. & KOÓS, J. 1997: Mezőkövesd-Mocsolyás. In Raczky, P., Kovács, P. & Anders, A. (eds.), *Utak a múltba: az M3-as autópálya régészeti leletmentései* (Budapest, Magyar Nemzeti Múzeum and Eötvös Loránd Tudományegyetem Régészettudományi Intézet), 28–33.

KALICZ, N. & MAKKAY, J. 1977: *Die Linienbandkeramik in der grossen ungarischen Tiefebene* (Budapest, Akadémiai Kiadó).

KALICZ, N. & RACZKY, P. 1987: Berettyóújfalu-Herpály: a settlement of the Herpály culture. In Raczky, P. (ed.), *The Late Neolithic of the Tisza Region* (Budapest and Szolnok, Szolnok Museum), 113–34.

KELLY, R.C. 1993: *Constructing Inequality: the Fabrication of a Hierarchy of Virtue among the Etoro* (Ann Arbor, University of Michigan Press).

KEMP, B. 1989: *Ancient Egypt: Anatomy of a Civilization* (London, Routledge).

KRISTIANSEN, K. 1998: *Europe before History* (Cambridge, Cambridge University Press).

KUPER, A. 1988: *The Invention of Primitive Society: Transformations of an Illusion* (London, Routledge).

LATOUR, B. 1993: *We Have Never Been Modern* (New York, Harvester Wheatsheaf).

LEICK, G. 1999: *Who's Who in the Ancient Near East* (London, Routledge).

MASCHIO, T. 1998: The narrative and counter-narrative of the gift: emotional dimensions of ceremonial exchange in southwestern New Britain. *Journal of the Royal Anthropological Institute* 4, 83–100.

MEAD, M. 1943: *Coming of Age in Samoa: a Study of Adolescence and Sex in Primitive Societies* (Harmondsworth, Penguin; first published 1928).

MEIGS, A. 1990: Multiple gender ideologies and statuses. In Sanday, P.G. & Goodenough, R.G. (eds.), *Beyond the Second Sex: New Directions in the Anthropology of Gender* (Philadelphia, University of Pennsylvania Press), 99–112.

MEISENHEIMER, M. 1989: *Das Totenritual, geprägt durch Jenseitsvorstellungen und Gesellschaftsrealität: Theorie des Totenrituals eines kupferzeitlichen Friedhofs zu Tiszapolgár-Basatanya (Ungarn)* (Oxford, British Archaeological Reports).

MOMADAY, N.S. 1976: Native American attitudes to the environment. In Capps, W.H. (ed.), *Seeing with a Native Eye: Essays on Native American Religion* (New York, Harper & Row), 79–85.

MOORE, H.L. 1988: *Feminism and Anthropology* (Cambridge, Polity Press).

NEUGEBAUER, J.W. & NEUGEBAUER, C. 1997: *Franzhausen: das frühbronzezeitliche Gräberfeld I* (Vienna, Ferdinand Berger & Söhne).

O'SHEA, J. 1996: *Villagers of the Maros: a Portrait of an Early Bronze Age Society* (New York, Plenum Press).

PARKER PEARSON, M. 1999: *The Archaeology of Death and Burial* (Stroud, Sutton).

POSTGATE, N. 1992: *Early Mesopotamia: Society and Economy at the Dawn of History* (London, Routledge).

RACZKY, P. (ed.) 1987: *The Late Neolithic of the Tisza Region* (Budapest and Szolnok, Szolnok Museum).

RACZKY, P., ANDERS, A., NAGY, E., KURUCZ, K., HAJDÚ, Z. & MEIER-ARENDT, W. 1997a: Polgár-Csöszhalom-dülö. In Raczky, P., Kovács, T. & Anders, A. (eds.), *Utak a múltba: az M3-as autópálya régészeti leletmentései* (Budapest, Magyar Nemzeti Múzeum and Eötvös Loránd Tudományegyetem Régészettudományi Intézet), 34–43.

RACZKY, P., ANDERS, A., NAGY, E., KRIVECZKY, B., HAJDÚ, Z. & SZALAI, T. 1997b: Polgár-Nagy Kasziba. In Raczky, P., Kovács, T. & Anders, A. (eds.), *Utak a múltba: az M3-as autópálya régészeti leletmentései* (Budapest, Magyar Nemzeti Múzeum and Eötvös Loránd Tudományegyetem Régészettudományi Intézet), 47–50.

RACZKY, P., MEIER-ARENDT, W., KURUCZ, K., HAJDÚ, Z. & SZIKORA, A. 1994: A Late Neolithic settlement in the Upper Tisza region and its cultural connections (preliminary report). *A Nyíregyházi Jósa András Múzeum Évkönyve* 36, 231–6.

RADOVANOVIĆ, I. 1996a: *The Iron Gates Mesolithic* (Ann Arbor, International Monographs in Prehistory).

RADOVANOVIĆ, I. 1996b: Some aspects of burial procedure in the Iron Gates Mesolithic and implications of their meaning. *Starinar* 47, 9–20.

RADOVANOVIĆ, I. 1997: The Lepenski Vir culture: a contribution to interpretation of its ideological aspects. In Lasić, M. (ed.), *ANTIΔΩPON Dragoslavo Srejović* (Belgrade, University of Belgrade, Faculty of Philosophy, Centre for Archaeological Research), 85–93.

RAPPORT, N. 1996: Individualism. In Barnard, A. & Spencer, J. (eds.), *Encyclopaedia of Social and Cultural Anthropology* (London, Routledge), 298–302.

RAPPORT, N. 1997: *Transcendent Individual: Towards a Literary and Liberal Anthropology* (London, Routledge).

REGA, E. 1997: Age, gender and biological reality in the Early Bronze Age at Mokrin. In Moore, J. & Scott, E. (eds.), *Invisible People and Processes: Writing Gender and Childhood into European Archaeology* (London, Leicester University Press), 229–47.

RENFREW, C. 1973: Monuments, mobilization and social organization in Neolithic Wessex. In Renfrew, C. (ed.), *The Explanation of Culture Change: Models in Prehistory* (London, Duckworth), 539–58.

RIVIÈRE, P. 1995: Houses, places and people: community and continuity in Guiana. In Carsten, J. & Hugh-Jones, S. (eds.), *About the House: Lévi-Strauss and Beyond* (Cambridge, Cambridge University Press), 189–205.

ROBB, J. 1994: Gender contradictions, moral coalitions, and inequality in prehistoric Italy. *Journal of European Archaeology* 2, 20–49.

RÖSCH, M. 1993: Prehistoric land-use as recorded in a lake-shore core at Lake Constance. *Vegetation History and Archaeobotany* 2, 213–32.

SAHLINS, M. 1999: Two or three things that I know about culture. *Journal of the Royal Anthropological Institute* 5, 399–421.

SANTOS-GRANERO, F. 1991: *The Power of Love: the Moral Use of Knowledge among the Amuesha of Central Peru* (London and Atlantic Highlands, NJ, Athlone Press).

SHANKS, M. 1999: *Art and the Early Greek State: an Interpretive Archaeology* (Cambridge, Cambridge University Press).

SHANKS, M. & TILLEY, C. 1987: *Social Theory and Archaeology* (Cambridge, Polity Press).

SHENNAN, S.E. 1975: The social organization at Branč. *Antiquity* 49, 279–88.

SHENNAN, S.E. 1982: From minimal to moderate ranking. In Renfrew, C. & Shennan, S. (eds.), *Ranking, Resource and Exchange* (Cambridge, Cambridge University Press), 27–32.

SHENNAN, S.J. 1986: Central Europe in the third millennium BC: an evolutionary trajectory for the beginning of the Bronze Age. *Journal of Anthropological Archaeology* 5, 115–46.

SHENNAN, S.J. 1993: Settlement and social change in central Europe, 3500–1500 BC. *Journal of World Prehistory* 7, 121–61.

SHENNAN, S.J. 1994: Grounding arguments about burials. *Archaeological Dialogues* 1, 124–5.

SHERRATT, A. 1981: Plough and pastoralism: aspects of the secondary products revolution. In Hodder, I., Isaac, G. & Hammond, N. (eds.), *Pattern of the Past: Studies in Honour of David Clarke* (Cambridge, Cambridge University Press), 261–305.

SHERRATT, A. 1982: Mobile resources: settlement and exchange in early agricultural Europe. In Renfrew, C. & Shennan, S. (eds.), *Ranking, Resource and Exchange* (Cambridge, Cambridge University Press), 13–26.

SHERRATT, A. 1987: Cups that cheered. In Waldren, W.H. & Kennard, R.C. (eds.), *Bell Beakers of the Western Mediterranean* (Oxford, British Archaeological Reports), 81–114.

SHERRATT, A. 1994: Core, periphery and margin: perspectives on the Bronze Age. In Mathers, C. & Stoddart, S. (eds.), *Development and Decline in the Mediterranean Bronze Age* (Sheffield, J.R. Collis Publications), 335–45.

SHERRATT, A. 1995. Reviving the grand narrative: archaeology and long-term change. *Journal of European Archaeology* 3, 1–32.

SOFAER DEREVENSKI, J. 1997: Age and gender at the site of Tiszapolgár-Basatanya, Hungary. *Antiquity* 71, 875–89.

SOFAER DEREVENSKI, J. 2000: Rings of life: the role of early metalwork in mediating the gendered life course. *World Archaeology* 31, 389–406.

SREJOVIĆ, D. 1972: *Europe's First Monumental Sculpture: New Discoveries at Lepenski Vir* (London, Thames & Hudson).

STEVANOVIĆ, M. 1997: The age of clay: the social dynamics of house destruction. *Journal of Anthropological Archaeology* 16, 334–95.

STRATHERN, A. 1971: *The Rope of Moka* (Cambridge, Cambridge University Press).

STRATHERN, A. & STEWART, P.J. 2000: Dangerous woods and perilous pearl shells: the fabricated politics of a longhouse in Pangia, Papua New Guinea. *Journal of Material Culture* 5, 69–89.

STRATHERN, M. 1987: Introduction. In Strathern, M. (ed.), *Dealing with Inequality: Analysing Gender Relations in Melanesia and Beyond* (Cambridge, Cambridge University Press), 1–32.

STRATHERN, M. 1988: *The Gender of the Gift: Problems with Women and Problems with Society in Melanesia* (Berkeley, University of California Press).

STRATHERN, M. 1992: Parts and wholes: refiguring relationships in a post-plural world. In Kuper, A. (ed.), *Conceptualizing Society* (London, Routledge), 75–104.

STRATHERN, M. 1996: Cutting the network. *Journal of the Royal Anthropological Institute* 2, 517–35.

TAYLOR, A.C. 1996: The soul's body and its states: an Amazonian perspective on being human. *Journal of the Royal Anthropological Institute* 2, 201–15.

THOMAS, J. 1996: *Time, Culture and Identity: an Interpretive Archaeology* (London, Routledge).

TILLEY, C. 1994: *A Phenomenology of Landscape: Places, Paths and Monuments* (Oxford, Berg).

TREHERNE, P. 1995: The warrior's beauty: the masculine body and self-identity in Bronze Age Europe. *Journal of European Archaeology* 3, 105–44.

TRINGHAM, R. 1991: Households with faces: the challenge of gender in prehistoric architectural remains. In Gero, J. & Conkey, M. (eds.), *Engendering Archaeology: Women and Prehistory* (Oxford, Blackwell), 93–131.

WEINER, J.F. 1991: *The Empty Place: Poetry, Space, and Being among the Foi of Papua New Guinea* (Bloomington and Indianapolis, Indiana University Press).

WHITTLE, A. 1996: *Europe in the Neolithic: the Creation of New Worlds* (Cambridge, Cambridge University Press).

WHITTLE, A. 1997: *Sacred Mound, Holy Rings. Silbury Hill and the West Kennet Palisade Enclosures: a Later Neolithic Complex in North Wiltshire* (Oxford, Oxbow).

WHITTLE, A. 1998a: Fish, faces and fingers: presences and symbolic identities in the Mesolithic–Neolithic transition in the Carpathian basin. *Documenta Praehistorica* 25, 133–50.

WHITTLE, A. 1998b: Beziehungen zwischen Individuum und Gruppe: Fragen zur Identität im Neolithikum der ungarischen Tiefebene. *Ethnographisch-Archäologische Zeitschrift* 39, 465–87.

The Birth of Architecture

RICHARD BRADLEY

SCALES OF ANALYSIS

THIS VOLUME CONTAINS TWO SETS OF papers, one from archaeologists and the other from a wider group of scholars with an interest in sociocultural evolution. The original meeting also included a number of social anthropologists. It may be helpful to locate this chapter in relation to these different fields.

Evolutionary biologists are concerned with the origins of social institutions at a very general level, whilst anthropologists are more interested in the content of specific institutions. Archaeologists occupy a middle ground. On the one hand, they have access to the extended timescale that social anthropologists lack, but they can also provide some of the detail that is not available to the biologist. For that reason they have a choice of two different perspectives.

They have another choice, too. They may use their distinctive data to 'diagnose' the general character of social institutions in the past or, like the anthropologist, they may prefer to study the details of particular situations. Here it is possible to interpret specific practices and the mechanisms by which they were established and maintained. The perspectives of anthropology and evolutionary biology are by no means incompatible, and practitioners of each discipline may study the same phenomena at different scales (Harrison & Morphy 1988). Archaeology is unusual because its distinctive material allows researchers to move between these two approaches or even to apply them to the same subject matter.

MONUMENTS AND SOCIAL EVOLUTION

Monumental architecture is central to each of these agendas, since it can be studied on either scale of analysis. This chapter attempts to do just that. On a general level there seems little doubt that monument building developed alongside sedentisim and that in most cases it was associated with farming rather than hunting and gathering. At the same time, it is a feature that has often been

Proceedings of the British Academy, **110**, 69–92, © The British Academy 2001.

used by archaeologists as a way of studying the emergence and operation of particular institutions.

The title of this chapter, 'The Birth of Architecture', is also that of a poem by W.H. Auden which forms the prologue to the verse sequence that he called 'Thanksgiving for a Habitat' (Auden 1976: 687). In it he refers to a whole series of major buildings, extending from Classical Greece to Victorian England. They include the Acropolis, Chartres Cathedral, Blenheim, and the Albert Memorial, but Auden's account also includes Stonehenge. For him, these constructions represent a single phenomenon.

This suggests several observations which I shall develop in the course of this chapter. There is the title: the poem is about the origins of 'architecture'. Although many animals can build complex structures, they do not invest them with symbolic meanings (Ingold 1983). Yet symbols are very important in the examples quoted here. The buildings mentioned by Auden provide a broad sample of human achievements. Chartres Cathedral is a celebration of the beliefs of medieval Christianity and the Acropolis is the crowning work of the Athenian city-state. Blenheim and the Albert Memorial extol the achievements of two particular individuals, but, like the other examples, they also commemorate the political structures that gave them their authority. All these buildings make statements that can only be understood in relation to particular institutions.

But what of Auden's other examples? He refers to Stonehenge and also to the monuments that he calls 'gallery-graves', a term which is less often used by archaeologists today. How far would a similar interpretation be justified here? In the case of megalithic tombs, which are among the oldest monuments in Europe (Sherratt 1990), what inferences should be drawn from their first appearance?

This discussion follows Auden in regarding prehistoric monuments as kinds of architecture in their own right. They impose an artificial order on the use of space, they are often built on a massive scale, using enormous amounts of human labour, and they seem to have been constructed according to designs that were the expression of particular ideas about the world. They result from the co-ordinated energies of many people working together to achieve a common aim, and in most cases they were made of raw materials that were likely to survive for unusual lengths of time. Yet despite these specific qualities, they have a restricted distribution in the past. None seems to have been built until the advent of the use of domesticated resources, although there are many cases in which there is evidence to suggest they were first constructed before a farming economy was well established (Bradley 1998: Chapters 1–5).

Even so, there is a tension between the general and the particular. Discussing the birth of architecture, Auden says that certain buildings were the work of 'the same Old Man', but, he continues, we can see what the 'old man' did yet may not understand why that happened (Auden 1976: 687): even if we were to know the builder's own conception of these structures, we would be unable to account for the creation of monuments in the first place. There remains a need for high-level theory.

There have been three main approaches to the significance of monument building as a general phenomenon. The first is that of Bruce Trigger (1990) who uses what he terms a thermodynamic model. This makes use of two principles of general application. He employs the Principle of Least Effort (Zipf 1949) to identify constructions built on an extravagant scale, and at the same time he sees their creation as a form of conspicuous consumption. Taken together, these principles identify a widely occurring phenomenon in the archaeology of prehistoric societies:

> In human societies, the control of energy constitutes the most fundamental and universally recognized measure of political power. The most basic way in which power can be symbolically reinforced is through the conspicuous consumption of energy. Monumental architecture, as a highly visible and enduring form of such consumption, plays an important role in shaping the political and economic behaviour of human beings. (Trigger 1990: 128)

Monument building provides a medium for social display and an arena in which competition for authority is worked out, and, for Trigger, that is why it characterizes a formative stage in the evolution of political structures. This also explains why comparable evidence is found in so many different societies.

A similar approach has been suggested by John Cherry (1978). He describes the distinctive role of monument building in two situations. It accompanies the growth of social complexity and helps to bring people together in the creation of a common project whose symbolism may well provide a supernatural sanction for their activities. It assumes a similar significance if the political structure is threatened, in which case the construction of monuments may be one method by which the integration of society can be renewed. It is a model that Cherry has applied to the Minoan peak sanctuaries of Crete, but the same interpretation has been advanced for the changing scale and labour demands of Egyptian pyramids (Rathje 1975).

There is a certain tension between the title of Auden's poem and that of the sequence of which it forms a part. He talks about the 'birth of architecture', but architecture is a particular form of material culture that is peculiar to our own species. Yet the poem is one of a group entitled 'Thanksgiving for a Habitat'. The word 'habitat' refers to something wider than the built environment, for it describes the natural home of any organism. There is an important difference between something that is common to all animals and a feature which is only present among human beings, yet recently those two characteristics have been brought together in what its proponents call a Darwinian framework.

To some extent this is an elaboration of the position taken by Trigger, but with a more explicit emphasis on adaptation and reproductive success. An idea that has played a prominent part in archaeological studies of monuments is 'wasteful advertising' (Dunnell 1999). Again this has much in common with Trigger's thesis. The model treats the building of conspicuous structures as a form of display by which elites use human labour to signal their competitive abilities:

> Monumental architecture is 'wasteful' in the Darwinian sense because it represents an expenditure of energy and resources that might otherwise have been directed towards reproduction and maintenance of offspring. However, in [this] model monument construction is also 'smart' advertising since it benefits both the signaller and receiver, whether that receiver is a potential competitor or a potential follower. By paying the extravagant fitness costs of monument construction, an elite person signals his or her ability to compete in political contests which, in turn, determine access to resources and mates. (Aranyosi 1999: 357)

All the approaches that I have summarized share a similar problem. They identify a general phenomenon and explain it in terms of a theory of universal application, but in doing so, they make little use of the available evidence. That is not to say that we should abandon these attempts to link the archaeological record into a wider intellectual framework. Rather, they are *insufficient* because they do not do justice to the sheer diversity of the monuments that they are discussing. There is room for a second level of analysis which might be equally informative. Trigger acknowledges this when he says that he is 'not challenging the observation that in each early civilization temples, palaces, and tombs had highly idiosyncratic meanings, which were either read into, or determined, such features as their shape, orientation, decoration, colour, and the materials out of which they were constructed' (1990: 128–9). Instead of using monuments to document the origins of institutions in general, we could employ the detailed observations that archaeologists have made at these sites to suggest some of the differences between the people who built them.

In fact monument building has also been interpreted as one of the diagnostic features of a particular stage of social evolution. Again this approach depends on a high degree of generalization. Unlike anthropologists, prehistorians work with dead informants. Their only means of communication is through material things, and the dialogue is largely one-sided. For that reason archaeologists have looked for widely occurring features that may be shared with the ethnographic record in the hope of recognizing phenomena that could allow them to diagnose the character of extinct social systems. Thus in the 1960s Elman Service saw monument building as one of the diagnostic features of the societies that he classified as chiefdoms (Service 1962: 142–77). Along with a range of other characteristics, monument building could be recognized in the archaeological record, and this seemed to offer a clue to the nature of social organization in parts of prehistoric Europe. For Colin Renfrew (1974) it also seemed to identify a significant threshold in the development of particular communities in regions that extended from southern England to the Aegean. Edmund Leach (1973) criticized this approach when it was first suggested in archaeology, but, whatever its merits at the time, thirty years later it is clear that it did inspire a new generation of prehistorians to think more boldly about the past.

Again this is an approach which operates on a general level. In its original formulation the identification of chiefdoms in prehistoric Wessex depended on comparing the archaeological evidence from that region with a list of characteristic features that Service had compiled from ethnographic sources. The difficulty is that not all of these occurred in every case and that the range of societies that Service characterized as chiefdoms was exceptionally diverse. Moreover, his work was linked to an explicitly evolutionary hypothesis in which societies developed in a prescribed sequence and attained progressively greater levels of complexity. More recent work has either avoided discussing chiefdoms or has accepted that the term covers a number of different kinds of community (Earle 1991).

For our purposes one point is particularly important. Renfrew's interpretation of the prehistoric sequence in Wessex (Renfrew 1973) — one which he later applied in modified form to Neolithic Orkney (Renfrew 1979) — was explicitly based on the evidence of the monuments that were built there, and in this respect it anticipated the generalizing models considered so far. It discussed the changing political geography of his study areas in relation to the size and spacing of successive forms of public monument and the amounts of human labour that were required to build them. Some of these labour estimates have been modified in more recent work, and in a few cases fieldwork has even changed the sequence in which particular monuments were built (Bradley 1991), but none of these developments affects the power of the basic argument. In his interpretation the study of monuments plays a central role in the investigation of social institutions.

Such a bold interpretation may be less popular today, but perhaps this is only because generalizing models are less fashionable in archaeology. Through a growing *rapprochement* with social anthropology, archaeologists working in Europe have begun to appreciate the possibilities of a much closer reading of extinct material culture. That is only possible because they have accepted that material culture itself is a medium that can be used strategically. It conveys information in rather the same manner as spoken language and it has the same capacity to be interpreted in different ways. Indeed, it is possible to talk, as Tilley (1999) has recently done, in terms of 'solid metaphors'. This term refers to the complex networks of interconnections that can be identified between different media in the past, such as house plans, the layout of monuments, the places in which artefacts were deposited, and the wider organization of the landscape. This provides a less rigid framework than the structuralism by which it is influenced (Hodder 1982), but it also requires more evidence than archaeology can often provide. Fortunately, monumental architecture is one of the fields in which that requirement is met. In the right circumstances — where the available material is well preserved and well recorded — it may be possible to consider some of these buildings in their local settings. In doing so, it may also be possible to explore the many different ways in which their creation and use can shed light on the characteristics of early social institutions. We must accept that the greater the detail in which specific monuments are investigated, the more difficult it will become to provide a single, clear-cut interpretation. There is an inevitable subjectivity here, and the best way of assessing these ideas is to consider how many of the observations can be accommodated by a single interpretative scheme and how many others remain outside it.

One further qualification is needed. This chapter considers how early monuments were connected with social institutions in prehistoric Britain and Ireland. It is limited to these areas for two reasons. First, these are among the places in which generalizing approaches have already been employed. The Darwinian 'wasteful advertizing' model has been applied by Aranyosi (1999) to the Irish Neolithic, whilst the same period in both Wessex and Orkney was studied by Colin Renfrew in some of the first applications of the chiefdom model in European archaeology (Renfrew 1973, 1979).

The other reason for limiting this account to British and Irish prehistory is because specialized kinds of monuments are a particular feature of northern and western Europe, rather than areas further to the south and east, where the main focus for symbolic elaboration was on the settlement and the house. Although that evidence has an obvious importance here, the references made by early monuments in Britain and Ireland are altogether wider, and for that reason they are less easy to explain in practical terms. In order to provide a simple narrative, the case studies are considered in chronological order.

EARLIER NEOLITHIC MONUMENTS

Although this account considers a part of Europe in which settlements and houses were not the main focus of symbolic elaboration, both provide the background to the first of these studies.

This concerns the Neolithic period, the phase in which domesticated resources were adopted in Britain and Ireland. Although it is fashionable to play down the impact of long-distance contacts as the main source of change, in this case it is hard to do so. The main domesticated resources associated with early farming had to be introduced from overseas, and there was little precedent for Neolithic material culture among the existing inhabitants of these islands (Thomas 1999). Still less was there an indigenous background to the forms of monumental architecture that developed at this time. Not only were they closely related to prototypes on the continent, but it seems as if their distinctive symbolism could only be understood in relation to their history on the European mainland (Bradley 1998: Chapters 3–5). That is because their characteristic forms referred back to settlements and houses of types that had already gone out of use generations earlier. These monuments enshrined the memory of an ancestral way of life that no longer conformed to reality.

The agricultural settlement of large parts of central and western Europe is epitomized by the Linear Pottery Culture and its successors (Whittle 1996: Chapter 6). The settlements of these groups maintained a strong uniformity across time and space, extending between about 5300 BC and 4500 BC and maintaining the same basic settlement pattern over an area reaching from Bohemia to Poland and from Austria to northern France, although there is evidence that this cohesive structure was breaking down towards the end of the sequence. Their settlements are typified by massive longhouses. The earliest of these buildings were strictly rectangular, whilst some of the later examples had a tapering ground plan and were broader towards one end. These houses were often found in groups. The individual buildings shared the same alignment, they were widely spaced, and only rarely were they rebuilt in the same positions. Some of the later settlements were associated with ditched enclosures. In some cases these earthworks contained the living area, whilst in others they were not built until the settlement had been abandoned.

After about 4500 BC longhouses went out of use and settlements became more ephemeral, perhaps suggesting a greater emphasis on mobility (Thomas 1999: Chapter 2). Domestic buildings seem to have taken a less massive form. The enormous domestic structures of earlier generations seem to have been replaced by a series of equally massive *monuments*. It is at this stage, towards 4000 BC, that Neolithic material culture first appears in Britain and Ireland, and it is here that we encounter stone and earthwork monuments without any of the domestic buildings that provided their source of inspiration (Bradley 1998: Chapter 1).

It is the symbolism of these early monuments that needs emphasizing now. There are two kinds of structures to consider. There are elongated mounds or cairns, which are associated with the remains of the dead. These features were often the outcome of a prolonged sequence of activity and some may have replaced small buildings or enclosures where human remains underwent a series of transformations before the unfleshed bones were arranged in their final configuration (Barrett 1994: 54–65). It has often been observed that the form of the mounds and cairns is very similar to that of continental longhouses (Bradley 1998: Chapter 3), and this case is greatly strengthened by recent work in northern France which shows that similar structures were created on top of

buildings of this type (Mordant 1997). Although this happened some time
after those buildings had gone out of use, the fact that the mounds shared the
dimensions and orientations of the longhouses suggests that a memory
remained of the original configuration of the settlement. Similar mounds were
built over a lengthy period of time, but after that formative period was over
their prototypes were no longer constructed. That is the situation that we find
in Britain and Ireland, where the size, layout, and even the orientation of these
monuments recall the form of domestic buildings, of a type which does not
occur in these islands. Where Neolithic houses are found they are smaller and
lighter structures (Darvill & Thomas 1996).

The same argument applies to the earthwork enclosures, whose ditches are
usually interrupted by a number of causeways, creating the impression of a
series of elongated pits. Again they are first found in Britain around 4000 BC,
although this kind of monument originated on the continent whilst longhouses
were still in use. At first they were associated with groups of domestic buildings
and sometimes they enclosed entire villages. Very similar enclosures were still
constructed after the settlement pattern had changed, and by this stage very
few of them seem to have been used in daily life (Bradley 1998: Chapter 5).
Rather, they became ceremonial centres, where the main archaeological
evidence is for feasting, animal sacrifice, the deposition of rare and exotic
artefacts, and the treatment of the dead. It seems as if the form of the earthwork perimeter evoked the idea of a settlement, but a settlement of a type that
no longer existed in reality. Again the earthwork was modelled on a prototype
that had been current in the past.

The same interpretation would apply to the British sites (such earthworks
are uncommon in Ireland). For the most part these enclosures appear to have
been built on the margins of the settled landscape, sometimes in small clearings
in the forest, but the mortuary mounds or 'long barrows' were rarely far away
and can be found in unusually high numbers near to these monuments
(Thomas 1999: Chapter 6). Figure 1 illustrates one such landscape in the Great
Ouse Valley, together with outline plans of two excavated monuments on the
Fen edge. Nowhere in this landscape is there clear evidence of residential buildings and in most areas (again Ireland is something of an exception) the remains
of houses are difficult to find. Those that are known exist in virtual isolation. In
many areas all that survive of the settlements are scatters of artefacts, or shallow pits dug into the subsoil. The discovery of specialized monuments contrasts with the evidence of daily life.

So much research has been devoted to investigating the Neolithic landscape
that it no longer seems likely that the bias is due to the work of archaeologists.
Rather, the elaboration of monuments seems to have taken place at the expense
of the settlements of this period. Why was this the case?

It is by no means obvious why Neolithic culture was adopted in Britain and

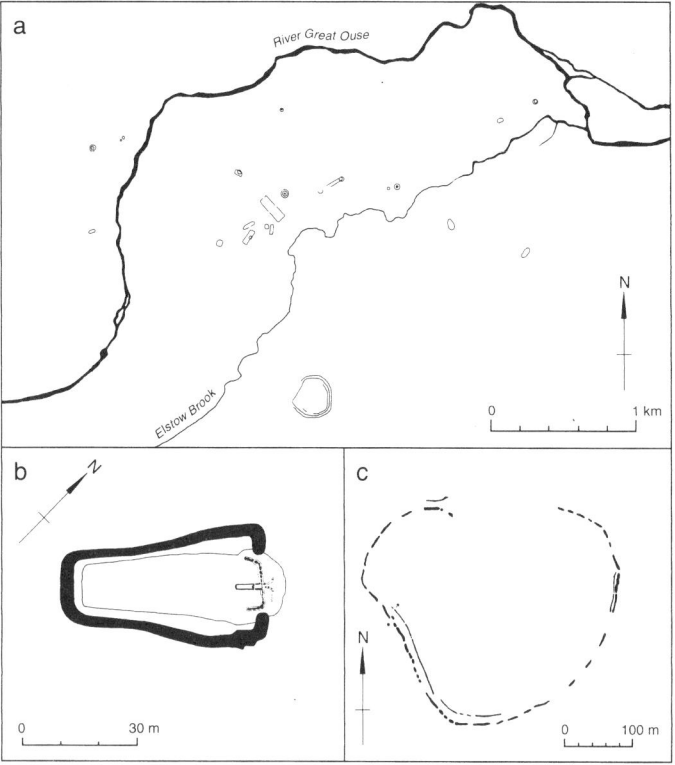

Figure 1. a: Earlier Neolithic monuments in the Great Ouse Valley (after Dawson 1996); b: outline plan of the Haddenham long barrow (after Hodder & Shand 1988); c: outline plan of the Haddenham causewayed enclosure (after Hodder 1992). Drawing: Steve Allen.

Ireland, nor is it clear how far the change was due to settlement from overseas, but this process did not take place in isolation. Something very similar happened at the same time in southern Scandinavia (Whittle 1996: Chapter 7). This raises an important point. Whatever the geographical sources of the local Neolithic, they are likely to have been extremely diverse and we can recognize individual points of resemblance between artefacts or monuments in these islands and those occurring across a vast area extending from Brittany to Denmark. That is not surprising considering the geographical position of Britain and Ireland in relation to the European landmass, but it does mean that links may have existed between particular parts of the study area and regions of the mainland that had fewer contacts among themselves. As a result, the British and Irish Neolithic has a distinctive identity of its own and does not reproduce the material culture of any one area of continental Europe.

Under these circumstances how would it be possible to create a sense of identity, and how might this have led to the development of new social institutions? Perhaps the creation and use of monuments had an important role to play and both the long mounds and earthwork enclosures were built as part of that process. That is because in their different ways both recalled an ancestral way of life on the European mainland: one which was dominated by a sense of community. They referred to the ancestors who may have lived together in the longhouses and to the separate households who inhabited the domestic enclosures. That no longer reflected reality by the time that Britain and Ireland adopted a Neolithic way of life, but the crucial transformation had already begun before that time. The landscapes of the living — landscapes which contained very few specialized structures — had already been replaced by landscapes of the dead in which that ancestral way of life was represented by monuments of kinds whose symbolism could only be explained by reference to the past. If so, then these were *landscapes of memory*, and it was through sharing in a similar origin myth that the people of the insular Neolithic were able to create their own sense of community. It matters very little how many of them were settlers from overseas and how many belonged to the indigenous population. What is important is that they subscribed to an origin myth and devoted themselves to its promulgation through the work of monument building. In constructing long barrows and enclosures they were, quite literally, helping to construct their own institutions, and they expressed their commitment to those ideas though buildings that would last for generations.

LATER NEOLITHIC MONUMENTS

If those structures referred back to a continental homeland, real or imagined, the next major group of monuments to be built were more closely integrated into their immediate surroundings. For the most part they were buildings of types that developed within Britain and Ireland. By the Later Neolithic period — that is to say, from about 3000–2500 BC — long barrows and causewayed enclosures had been succeeded by a fresh generation of monuments. There is no continuity between these successive forms of structure, and the new kinds of architecture — passage graves, stone circles, and the earthworks known as 'henges' — may be closely related to developments in the north and west where causewayed enclosures were never common. Later Neolithic monuments were widely distributed across a landscape quite large parts of which had been cleared and settled by this time (Thomas 1999: Chapter 3). The artefacts found at these monuments sometimes originated in distant areas, suggesting a much wider range of contacts within Britain and Ireland than had existed before.

During the earlier part of the Neolithic the main frame of reference of

stone and earthwork monuments seems to have been the houses and settlements of an ancestral homeland. Now the dominant symbolism of the larger
monuments united the dwellings of the living with the landscapes in which they
were built. Instead of a series of rigidly demarcated monument types, there was
more of a continuum between domestic buildings and ceremonial sites. This is
shown especially clearly by the archaeology of two different areas, Wessex and
Orkney. Both are regions which have been studied by Colin Renfrew (1973,
1979). In the case of Orkney this account also draws on the research of Colin
Richards (1993).

The situation in Orkney is especially relevant here. This is one of the very
few areas in which both tombs and houses are preserved, and archaeologists
have tended to treat its evidence as something unique. Its state of preservation
is certainly unprecedented, but there is reason to think that the relationships
between these different elements are reflected in other regions. So is the setting
of the monuments in the landscape.

There are three elements to consider here and, whilst their chronology does
pose certain problems, their histories most probably overlapped around 3000
BC. There are the stone-built houses which are best known from Skara Brae and
Barnhouse. These are generally found in small villages, and each individual
building had a stereotyped layout, with a roughly cruciform interior enclosed
by a circular outer wall. The main features of the living space were a stone-
built hearth, a 'dresser' against the rear of the structure facing the door, and a
series of recesses set into the thickness of the wall (Richards 1993). The tombs
of the same period were organized around a rather similar division of space,
although the central chambers were buried beneath a considerable mound or
cairn and could only be approached by a low entrance passage — hence their
description as 'passage graves'. Again there was a main chamber with a series
of smaller chambers or cells radiating from it, each of them approached
though a narrow entrance of its own (ibid.). Some of the decorated pottery
associated with these tombs was of the kind found in the settlements and recent
work has shown that both groups of structures were decorated with the same
kinds of abstract motifs (Bradley *et al.* 2001) (Figure 2). The passage graves
were generally located close to living areas.

Near the greatest tomb in Orkney, Maes Howe, there are two other sites.
One is the Later Neolithic settlement of Barnhouse, whilst the other is a
ditched enclosure, the Stones of Stenness, which contains a setting of enormous monoliths. Again they are likely to have been used over the same period.
The stone circle at Stenness employs uprights of exactly the same form as those
in the central chamber at Maes Howe, and at the centre of the ring of monoliths
there is an enormous slab-lined hearth just like those found in the domestic
buildings at Barnhouse (Richards 1993). In short, the enclosure, the
settlement, and the tomb conform to the same principles of order and there are

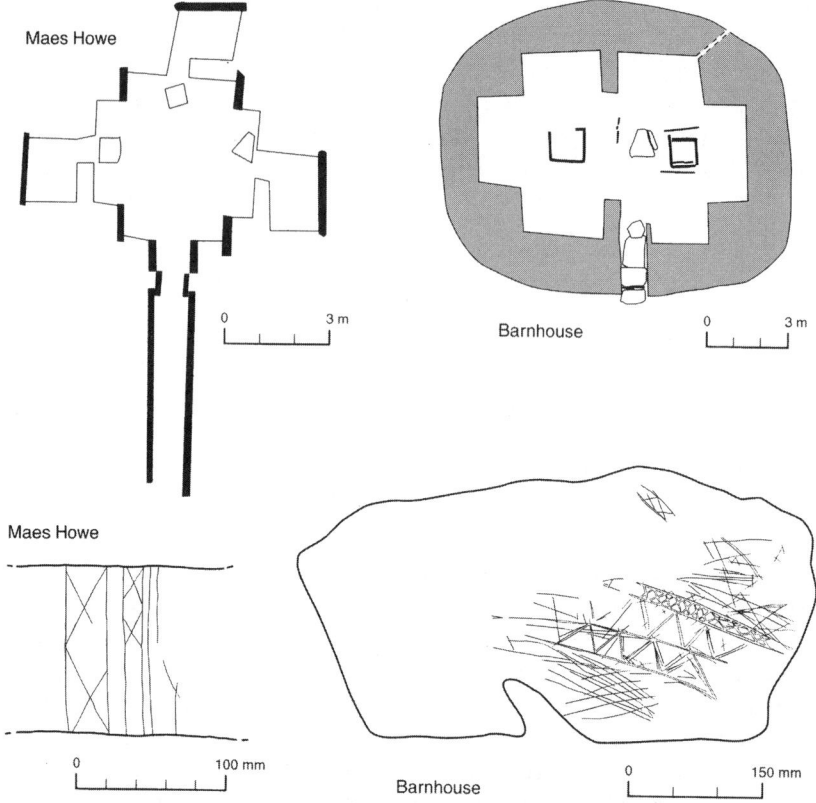

Figure 2. Simplified plans of the chamber at Maes Howe and a Grooved Ware house at Barnhouse, together with outline drawings of decorated stones from these two sites (after Bradley *et al.* 2001). Drawing: Steve Allen.

elements of material culture — ceramics and decorated stonework — which are shared between them.

But what of the landscape itself? In a recent paper Richards (1996) has suggested that the siting of the Stones of Stenness, and that of the Ring of Brodgar, a similar monument nearby, was very carefully chosen so that each of these circular enclosures appeared to be at the centre of a more extensive circular landscape, whose outer limit was marked by a horizon of hills: these sites were essentially arenas. The lochs in the foreground were equivalent to the enclosure ditches, which may have held water, whilst the outer bank or wall was the counterpart of the more distant barrier of high ground. Each monument could have epitomized the properties of a much wider area, and the houses, the stone settings, and the tombs could all have referred to this connection.

In other parts of Britain, and especially in Wessex, there is a similar overlap between domestic and public buildings, although this has hardly been acknowledged (Darvill & Thomas 1996). Again the principal sites included circular earthwork enclosures which in this case contained settings of large upright posts. There was a similar division of space inside the simple round-houses of this period. These can be compared directly with the timber circles associated with henges and even with the ground plan of the enclosures themselves (Figure 3). The argument can be taken further, as there are henges with stone settings in their centre which seem to represent monumentalized versions of the domestic hearth. The continuum even extends to the situation of these earthworks in the landscape. Thus the enormous henge monument of Durrington Walls is located inside a dry valley so that the impression created by the earthwork boundary is reinforced by the local topography. Other monuments, such as Avebury, occupy the middle of a large natural basin, so that the distant horizon echoes the form of the perimeter earthwork (Bradley 1998: 119–28). The very existence of this continuum suggests that there was no clear-cut distinction between the ritual and domestic worlds.

The artefact assemblages recovered from such sites extend along a similar continuum, from quite straightforward deposits to others with a more specialized character, identified by their distinctive composition and the manner in which they had been committed to the ground (Thomas 1999: Chapter 4). Some of the more striking artefacts are found in settlement sites, again suggesting that any rigid division between the ritual and domestic spheres would be inappropriate. Rather, it seems as though there was a continuous range of variation. Structured deposits are most obvious when they are found within specialized monuments, but they also occur across the surrounding landscape. It seems to be true, however, that the most varied assemblages are evidenced near to ceremonial centres.

How can this evidence be understood? Perhaps such places were conceived as 'big houses': as public buildings which symbolized the unity of the social groups who built and used them. This is an idea that has already been suggested in the New World (De Boer 1997). Such structures were enormously enlarged versions of the ordinary dwellings of this period, yet their purely symbolic character is obvious from the way in which some of them were rebuilt as free-standing rings of monoliths; indeed, the lintelled structure at Stonehenge seems to be modelled on just such a prototype (Gibson 1998). On the other hand, that does not account for the choice of a circular ground plan for so many different monuments. Perhaps that developed because those places were also perceived as *microcosms of the landscapes in which they were built*. Where the Earlier Neolithic world had been constructed around an origin myth, during this period domestic and ceremonial spaces were the mirror images of one another.

The recreation of the timber settings in stone is especially revealing here.

Richard Bradley

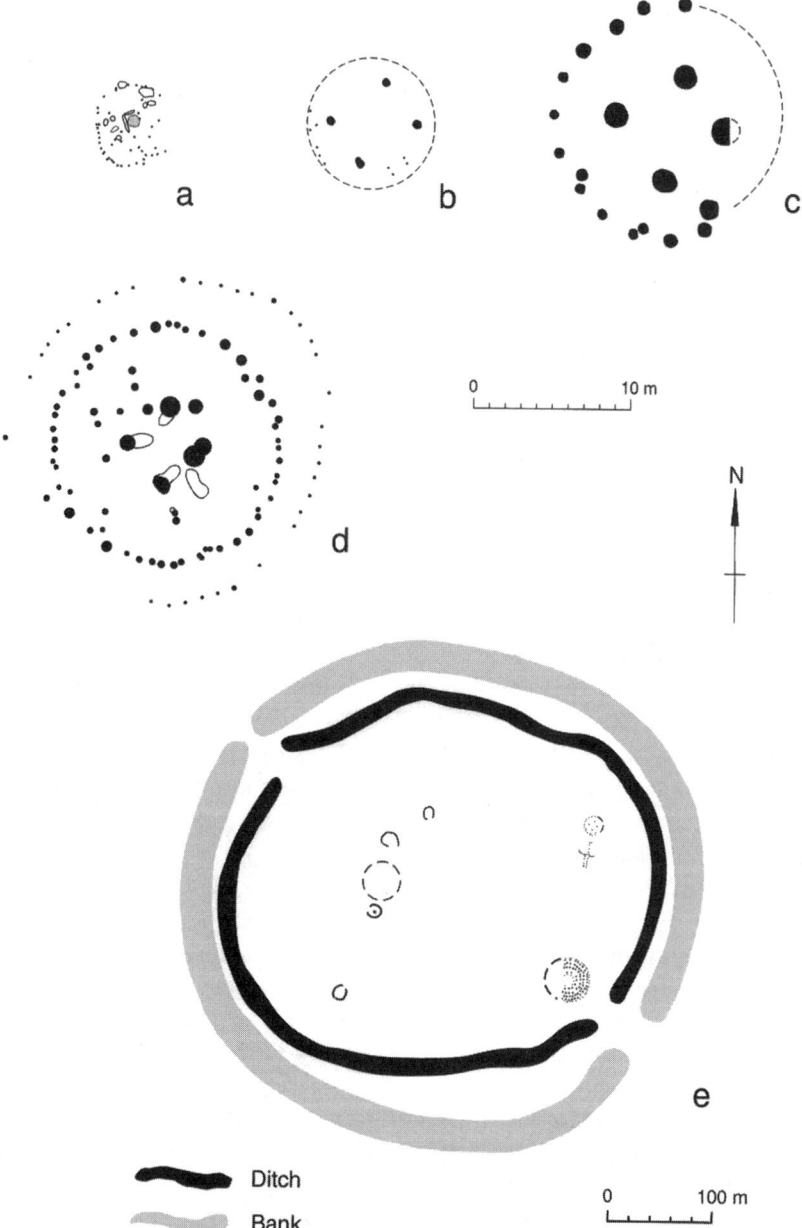

Figure 3. Later Neolithic houses and ceremonial monuments. a: Trelystan (after Britnell 1982); b: Wyke Down (after Green 1997); c: Durrington Walls (after Wainwright and Longworth 1971); d: Machrie Moor (after Haggarty 1991); e: the Durrington Walls henge (after Wainwright and Longworth 1971 with additions). Drawing: Steve Allen.

This was often the last phase of reconstruction at the monuments. The wooden buildings on these sites were often associated with placed deposits of artefacts and with the remains of feasts. The stone settings, on the other hand, are only rarely associated with artefacts and in their later phases they may contain human burials (Parker Pearson & Ramilisonina 1998). This has led to the idea that the use of an organic raw material — wood — was associated with the living population whilst the reconstruction of these monuments using an inorganic material — stone — may have been associated with the timeless qualities of the dead. At all events, many monuments went through a similar cycle, and it resulted in the creation of structures which have been able to resist natural decay for over four thousand years. The erection of these monuments may have brought the population together in structures that were conceived as houses for an entire community and models of the surrounding world. The rebuilding of these constructions in stone could have given them an added authority.

EARLIER BRONZE AGE MONUMENTS

The last case study comes from the period that archaeologists call the Bronze Age. In fact there is no simple division between this phase and its predecessor, but for present purposes one observation is certainly important. Although older monuments continued to be used, sometimes on a considerable scale, there is little evidence that equally elaborate structures were still being built. Much smaller monuments were created instead, most of them associated with human burials. In place of the large arenas of the Later Neolithic there were circular mounds and cairns.

The next example comes from that transitional period. Towards the end of the Neolithic sequence in Orkney, a rather different development was taking place at a site on the northern mainland of Scotland. At Raigmore, a wooden house associated with the same style of pottery as Barnhouse and Skara Brae was replaced by a massive cairn, with its kerbstones graded in height towards the south-west (Simpson 1996) (Figure 4). This is interesting for two reasons, for not only was the house directly replaced by a stone monument, but in doing so the builders changed its orientation. The house of the living extended from north-west to south-east but the funerary monument that took its place was laid out on a new axis at right angles to the original design. A cremation burial had been placed in the centre of the house, but so many more were associated with the cairn that it is reasonable to describe it as a kind of cemetery. The rebuilding of Raigmore as a monument for the dead inverted the alignment associated with the living.

Between about 2500 and 1500 BC, small circular monuments to the dead

Richard Bradley

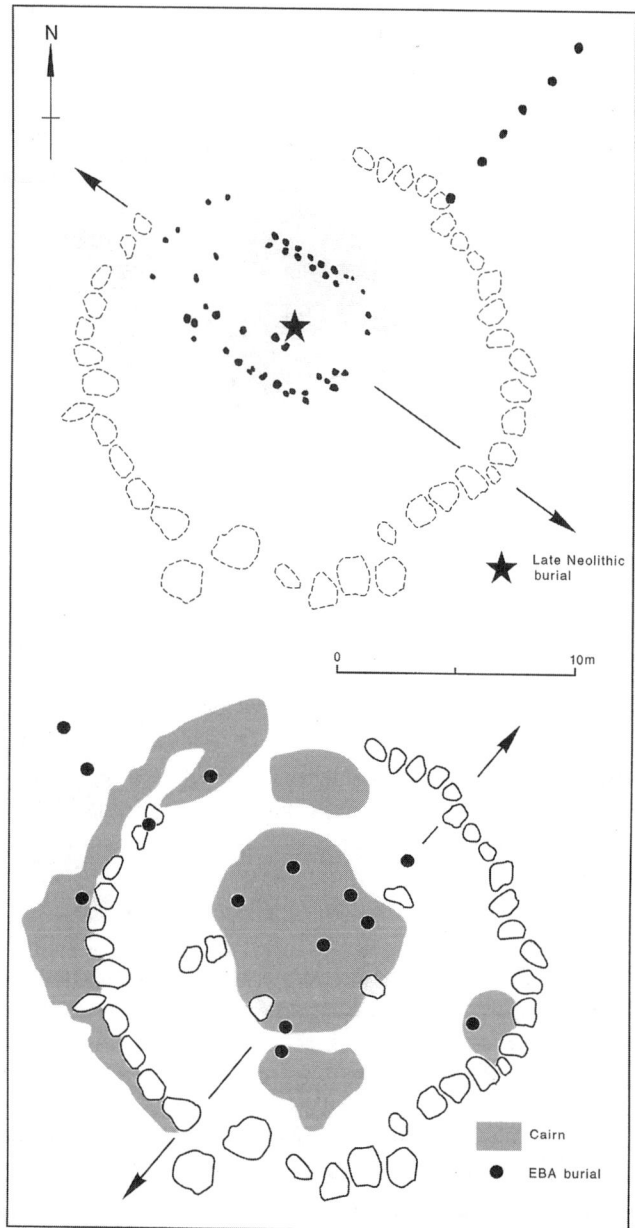

Figure 4. An interpretation of the archaeological sequence at Raigmore (after Simpson 1996). In the upper plan the postholes of the timber structures are shown in black and the outer kerb of the later cairn is indicated in broken outline. The lower plan shows the layout of the cairn that replaced the Neolithic house. The bold arrows indicate alignments of the successive buildings on the site. Drawing: Margaret Mathews.

came to dominate the landscape of Britain and Ireland (Barrett 1994: Chapter 5). Although they existed alongside the later use of henges and stone circles, the two sometimes coalesced and it was not unusual for graves to be dug inside earlier structures or even for burial cairns to be built there (Bradley 1998: Chapter 9). Although the mounds and cairns were once associated with the development of a 'single grave' tradition, this is really a misnomer as most well-excavated examples contain a number of separate deposits of human remains (Last 1998; Petersen 1972). Nevertheless it is true that they were often placed there on separate occasions and that each might have been accompanied by an appropriate selection of artefacts. At the same time, such monuments rarely existed in isolation and they frequently formed parts of larger cemeteries.

Although there are many variations, the common element is that the dead were buried within round mounds or cairns. Although these were enlarged as new burials were added, in their earlier phases they were often the same size and shape as the houses built during the same period, a connection that was only emphasized by embellishing some of these structures with rings of wooden posts (Ashbee 1960: Chapter 5). This is rather revealing, for there are further cases where burial mounds were built over the remains of domestic buildings (Lane 1984). The sequence seen at Raigmore is paralleled at other sites and raises the possibility that what archaeologists have thought of as burial mounds were conceived as houses for the dead.

There is a need for caution, as the domestic buildings of this period were never substantial structures and their remains are difficult to find. At the same time, there seems little doubt that the barrows and cairns might be built near to settlements. Underneath them there is often domestic refuse and some sites provide environmental evidence that they were constructed in farmland. That seems to be particularly true of the smaller burial mounds, whereas some of the most elaborate and richly furnished graves were covered by monuments built on higher ground (Peters 2000).

Many of these points can be illustrated by a cemetery at the Brenig, in North Wales (Lynch 1993) (Figure 5). This was on the spring line, in an area that had already been occupied during the Later Neolithic, but it is clear from environmental evidence that the main settled area was on lower ground some distance beyond the site itself. That was where cereals were being grown and it was from that area that people brought some of the turf used to construct the mounds. The land around the monuments, however, was mainly pasture. One of the Bronze Age cairns was built over the position of a wooden house, whilst there is evidence of another building and a group of domestic artefacts from the immediate area.

This example is among many sites where burial mounds were constructed in groups. Sometimes their layout was organized in formal patterns. They were

　　　　　　　　　　　　Richard Bradley

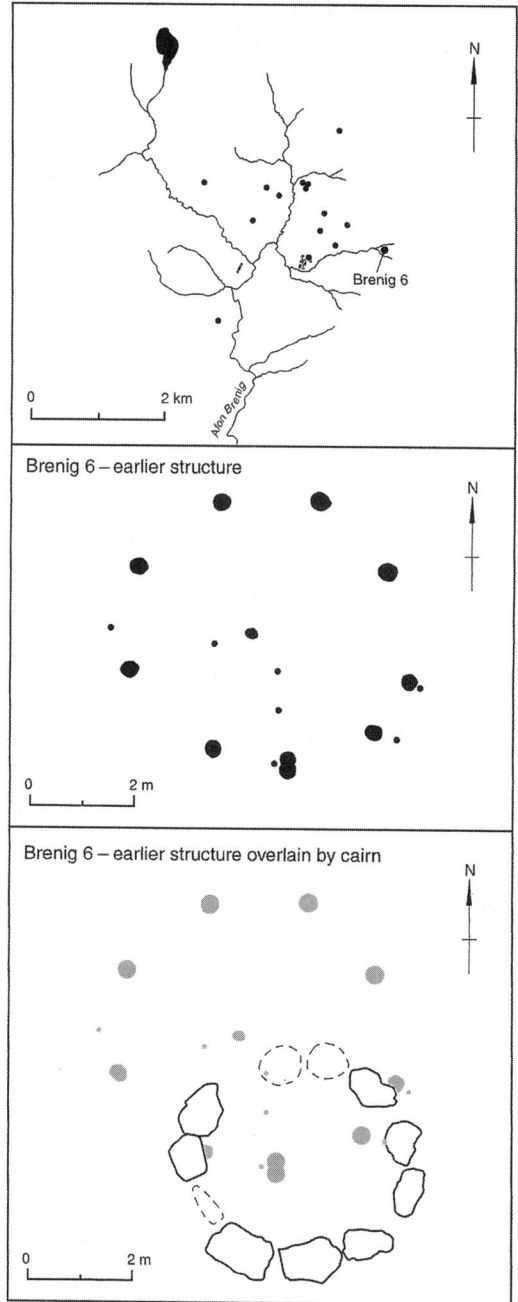

Figure 5. Outline plan of the Bronze Age cemetery at the Brenig (after Lynch 1993) with details of the timber building on Site 6 and the cairn overlying it. Drawing: Steve Allen.

frequently located in relation to older, Neolithic monuments, but in this particular case that does not seem to have happened. The cemetery contains a whole
array of small circular monuments, from complex burial mounds to stone
enclosures.

If single monuments were understood as representations of the house and
as lasting memorials to its occupants, then cemeteries of this kind must surely
be viewed as *the settlements of the dead*. That is why the spatial relationship
between these separate monuments could be so important (Mizoguchi 1992). If
the structural sequence at individual monuments reflected the changing history
of one social group, the linkages between different structures in the same cemetery — their juxtaposition, alignment, or even their avoidance — might also
provide some indications of the wider networks of alliance and obligation in
which those people were involved. It is not a new idea to claim that in the structure of such cemeteries we may be seeing the three-dimensional representation
of a genealogy (Barrett 1994: Chapter 5), but it is important to appreciate the
distinctive medium through which it was expressed. It may not be too much to
suggest that over the generations the relatively insubstantial dwellings of the
living were supplanted by the more massive structures of the dead.

It was only at the very end of this period that these priorities seem to have
changed and the nature of that change is revealing in itself. It represented something of a retreat from monument building, and it came at a time when we find
the first widespread evidence of productive mixed farming in Great Britain. At
just the point when one might have expected economic surpluses to be invested
in the creation of great public works, human energies seem to have been redi-
rected into the creation of more lasting settlements, land boundaries, and field
systems. The houses of the dead were replaced by the houses of the living
(Bradley 1998: Chapter 10).

It is from that period of change that the final illustration is taken. The last
burial mounds were built at the same time as field systems and land divisions
were becoming established. More substantial houses were also built at this time
and, in contrast to earlier practice, their positions seem to have been respected
by later generations so that their remains were not destroyed. The small burial
mounds of this phase were generally located close to the living area, and there
are cases in which their ground plans can be compared with one another.
Figure 6 shows one of the houses in the settlement at Itford Hill in Sussex and
compares it with the organization of the small cremation cemetery that served
that site (Holden 1972). There are many similarities. Both were circular and
both were about the same size. The house had a porch facing towards the south-
east and this arrangement is mirrored by a causeway in the ditch enclosing the
mound. It was on that side of the monument that most of the burials were
located. The resemblance even extends to the setting of wooden posts around
the edge of the mound. If there is any doubt about the closeness of this

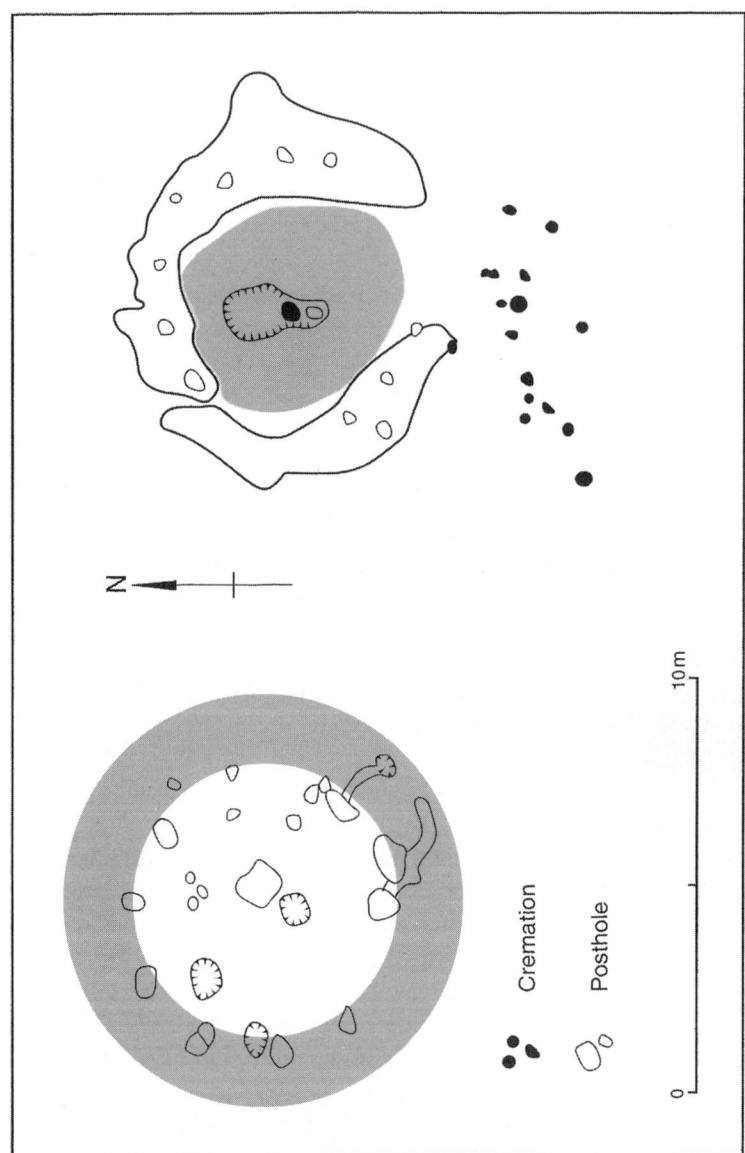

Figure 6. (Left) Outline plan of one of the Bronze Age houses at Itford Hill. The shaded area would have been under the eaves. (Right) Plan of the small barrow serving the settlement. The position of the mound is shaded and the extent of the enclosing ditch is shown in bold outline. Both structures share a rather similar organization of space. After Bradley (1998). Drawing: Margaret Mathews.

relationship, it is surely dispelled when we realize that one of the cremations in the cemetery was deposited in a broken pot, another fragment of which was discovered in the settlement.

SUMMING UP

At the beginning of this chapter I suggested that monument building was both a clue that archaeologists can use to the character of ancient social institutions and part of the process through which those institutions acquired and maintained their power. I have tried to illustrate that contention using three case studies from successive phases of British and Irish prehistory. In the first, I argued that the earliest farmers in Britain created a new sense of community by building monuments that were directly related to a mythical source of origin on the continent. By tracing the gradual development of those architectural forms from their invention some generations earlier, it was possible to suggest the meanings that they were intended to convey.

My second example came from a period when greater amounts of energy were devoted to monument construction and combined the results of research in Orkney with work in other parts of Britain. In each case it seemed as if the form of the largest monuments could only be understood in relation to a wider conceptual scheme. This extended from the layout of the individual house, though the configuration of a series of ceremonial centres, to the organization and perception of the landscape as a whole. The metaphor of the 'big house', as Warren De Boer (1997) has called it, seemed to permeate the archaeology of the Later Neolithic period and may have been one of the ways in which larger political structures were established and displayed. That is not to say that such monuments arose haphazardly — no doubt their development required both planning and direction — but it would be wrong to overlook their other aspect as projects that involved a considerable workforce in the execution of a common task.

Lastly, I considered how the burial mounds of the Earlier Bronze Age could be reinterpreted. So much research has been concerned with the contents of the various graves that it is hard to remember that any display of portable wealth would have been short-lived. After the funeral was over, it would survive only as a memory. The mounds that commemorated the dead, on the other hand, provided a tangible statement of the social order and a history of the local community. That may be why the houses of the dead had a longer currency than the dwellings of the living population.

All these interpretations, tentative as they are, could be subsumed within the general theories of monument building described in the first part of this chapter. They certainly do not contradict them, and for some researchers that

may be enough. The disadvantage is that the same theories account for observations that are so very different from one another when we study them in
detail, and it is those points of difference that are one of the strengths of the
archaeological record — as they are of ethnographic writing. But with this
predilection for a rather more fine-grained analysis there come inevitable dangers. There is more scope for differences of opinion, and an increased risk of
pure subjectivity. But I believe that both kinds of interpretation can be pursued
in tandem and that there is scope for profitable exchanges between them. It is
for individual scholars to decide on what seems to be the most appropriate scale
of analysis, and that may be partly a matter of personal inclination and partly
a response to the character of the data being studied. The doctrinal quarrels
that have characterized the study of prehistory for the past twenty years seem
to be largely over and it is time to move away from the introspection that they
engendered. There are many ways of studying the past and archaeology is only
one of them. Dispute is always less congenial than dialogue and it is in this
spirit that this chapter has been written.

REFERENCES

ARANYOSI, E.F. 1999: Wasteful advertising and variation reduction: Darwinian models for the
    significance of non utilitarian architecture. *Journal of Anthropological Archaeology* 18, 356–75.
ASHBEE, P. 1960: *The Bronze Age Round Barrow in Britain* (London, Phoenix House).
AUDEN, W.H. 1976: *Collected Poems* (London, Faber).
BARRETT, J. 1994: *Fragments from Antiquity* (Oxford, Blackwell).
BRADLEY, R. 1991: The pattern of change in British prehistory. In Earle, T. (ed.), *Chiefdoms:
    Power, Economy and Ideology* (Cambridge, Cambridge University Press), 44–70.
BRADLEY, R. 1998: *The Significance of Monuments* (London, Routledge).
BRADLEY, R., PHILLIPS, T., RICHARDS, C. & WEBB, M. 2001: Decorating the house of the
    dead: incised and pecked motifs in Orkney chambered tombs. *Cambridge Archaeological
    Journal* 11.
BRITNELL, W. 1982: The excavation of two barrows at Trelystan, Powys. *Proceedings of the
    Prehistoric Society* 48, 133–201.
CHAPMAN, J. 1999: Deliberate house-burning and the prehistory of central and eastern Europe.
    In Gustafsson, A. & Karlsson, H. (eds.), *Glyfer och arkeologiska rum — en vänbok till Jarl
    Nordbladh* (Gothenburg, Gothenburg University Institute of Archaeology), 113–26.
CHERRY, J. 1978: Generalization and the archaeology of the state. In Green, D., Haselgrove, C. &
    Spriggs, M. (eds.), *Social Organization and Settlement* (Oxford, British Archaeological Reports
    International Series 47), 411–37.
DARVILL, T. & THOMAS, J. (eds.) 1996: *Neolithic Houses in North-West Europe and Beyond*
    (Oxford, Oxbow).
DAWSON, M. 1996: Plantation Quarry, Willington. Excavations 1988–91. *Bedfordshire Archaeol-
    ogy* 22, 2–49.
DE BOER, W. 1997: Ceremonial centres from the Cayapas (Esmeraldas, Ecuador) to Chillicothe
    (Ohio, USA). *Cambridge Archaeological Journal* 7, 225–53.
DUNNELL, R. 1999: The concept of waste in evolutionary archaeology. *Journal of Anthropolog-
    ical Archaeology* 18, 243–50.

EARLE, T. 1991: The evolution of chiefdoms. In Earle, T. (ed.), *Chiefdoms: Power, Economy and Ideology* (Cambridge, Cambridge University Press), 1–15.

GIBSON, A. 1998: *Stonehenge and Timber Circles* (Stroud, Tempus).

GREEN, M. 1997: A second henge and Neolithic buildings uncovered on Wyke Down, Cranborne Chase, Dorset. *Past* 27, 1–3.

HAGGARTY, A. 1991: Machrie Moor, Arran. Recent excavations at two stone circles. *Proceedings of the Society of Antiquaries of Scotland* 121, 51–94.

HARRISON, G. & MORPHY, H. (eds.) 1998: *Human Adaptation* (Oxford, Berg).

HODDER, I. (ed.) 1982: *Symbolic and Structural Archaeology* (Cambridge, Cambridge University Press).

HODDER, I. 1990: *The Domestication of Europe* (Oxford, Blackwell).

HODDER, I. 1992: The Haddenham causewayed enclosure — a hermeneutic circle. In Hodder, I. (ed.), *Theory and Practice in Archaeology* (London, Routledge), 213–40.

HODDER, I. & SHAND, P. 1988: The Haddenham long barrow: an interim report. *Antiquity* 62, 349–53.

HOLDEN, E. 1972: A Bronze Age cemetery-barrow on Itford Hill, Beddingham, Sussex. *Sussex Archaeological Collections* 110, 70–117.

INGOLD, T. 1983: The architect and the bee: reflections on the work of animals and man. *Man* (new series) 18, 1–20.

LANE, P. 1984: Past practices in the ritual present: examples from the Welsh Bronze Age. *Archaeological Review from Cambridge* 5 (2), 181–92.

LAST, J. 1998: Books of life: biography and memory in a Bronze Age barrow. *Oxford Journal of Archaeology* 17, 43–53.

LEACH, E. 1973: Concluding address. In Renfrew, C. (ed.), *The Explanation of Culture Change* (London, Duckworth), 761–71.

LYNCH, F. 1993: *Excavations in the Brenig Valley* (Bangor, Cambrian Archaeological Association).

MIZOGUCHI, K. 1992: The historiography of a linear barrow cemetery. *Archaeological Review from Cambridge* 11.1, 39–49.

MORDANT, D. 1997: Le complexe des Réaudins à Balloy: enceinte et nécropole monumentale. In Constantin, C., Mordant, D. & Simonin, D. (eds.), *La culture de Cerny* (Nemours, L'Association pour la Promotion de la Recherche Archéologique en Ile-de-France), 449–79.

PARKER PEARSON, M. & RAMILISONINA. 1998: Stonehenge for the ancestors: the stones pass on the message. *Antiquity* 72, 308–26.

PETERS, F. 2000: Two traditions of Bronze Age burial in the Stonehenge landscape. *Oxford Journal of Archaeology* 19, 243–58.

PETERSEN, F. 1972: Traditions of multiple burial in Later Neolithic and Early Bronze Age England. *Archaeological Journal* 129, 22–55.

RATHJE, W. 1975: The last tango at Mayapan: a tentative trajectory of production-distribution systems. In Sabloff, J. & Lamberg-Karlovski, C. (eds.), *Ancient Civilization and Trade* (Albuquerque, University of New Mexico Press), 409–48.

RENFREW, C. 1973: Monuments, mobilization and social organization in Neolithic Wessex. In Renfrew, C. (ed.), *The Explanation of Culture Change* (London, Duckworth), 539–58.

RENFREW, C. 1974: Beyond a subsistence economy: the evolution of social organization in prehistoric Europe. In Moore, C.B. (ed.), *Reconstructing Complex Societies* (Ann Arbor, American School of Oriental Research), 69–95.

RENFREW, C. 1979: *Investigations in Orkney* (London, Society of Antiquaries).

RICHARDS, C. 1993: Monumental choreography, architecture and spatial representation in later Neolithic Orkney. In Tilley, C. (ed.), *Interpretative Archaeology* (Oxford, Berg), 443–81.

RICHARDS, C. 1996: Monuments as landscape. Creating the centre of the world in Neolithic Orkney. *World Archaeology* 28, 190–208.

SERVICE, E. 1962: *Primitive Social Organization: an Evolutionary Perspective* (New York, Random House).

SHERRATT, A. 1990: The genesis of megaliths: monumentality, ethnicity and social complexity in Neolithic north-west Europe. *World Archaeology* 22, 147–67.

SIMPSON, D. 1996: Excavation of a kerbed funerary monument at Stoneyfield, Raigmore, Inverness, Highland, 1972–3. *Proceedings of the Society of Antiquaries of Scotland* 126, 53–86.

THOMAS, J. 1999: *Understanding the Neolithic* (London, Routledge).

TILLEY, C. 1999: *Metaphor and Material Culture* (Oxford, Blackwell).

TRIGGER, B. 1990: Monumental architecture: a thermodynamic explanation of symbolic behaviour. *World Archaeology* 22, 119–32.

WAINWRIGHT, G. & LONGWORTH, I. 1971: *Durrington Walls. Excavations 1966–1968* (London, Society of Antiquaries).

WHITTLE, A. 1996: *Europe in the Neolithic: the Creation of New Worlds* (Cambridge, Cambridge University Press).

ZIPF, G.K. 1949: *Human Behaviour and the Principle of Least Effort* (Cambridge MA, Addison-Wesley).

Commodification and Institution in Group-Oriented and Individualizing Societies

COLIN RENFREW

THE SAPIENT PARADOX AND THE EMERGENCE OF MIND

IN THIS CHAPTER I HOPE TO touch on some problems and questions which I feel set the question of the emergence of 'mind' in a rather different light from that at present widely accepted. It seems a paradox (Renfrew 1996) that while the most significant steps in human evolution in the physical sense occurred more than 40,000 years ago, with the emergence of our species *Homo sapiens sapiens*, the salient aspects of human behaviour which distinguish our species so markedly from that of the other mammals emerged in many cases very much later. 'By their works ye shall know them' seems a good motto for the archaeol-ogist, and the most prominent of those works post-date the Upper Palaeolithic period. Yet there seems little doubt that the 'hardware', the human body and the brain, had attained its definite structure by that early date. Recent DNA studies do suggest that the 'out of Africa' view for the origins of our own species is likely to be correct, and that the African origin took place more than 100,000 years ago. So far as Europe is concerned, our species made its appear-ance some 40,000 years ago. It is becoming increasingly clear that, while there were probably many migratory episodes involved, the genetic differences within our species were and are quite limited. That genetic variability among humans at the present time has been well studied, and the differences do not seem very great.

When I have suggested in lectures to academic audiences that in the 30,000 years following the *sapiens* entry into Europe nothing very much of interest happened, Palaeolithic archaeologists have very rightly indicated that this is an exaggeration. They have pointed to a number of technical advances which took place during the Upper Palaeolithic period. One of the most notable was the development of figurative painting in the caves of southern France and northern Spain, along with the carving on bone and stone — the mobiliary art

Proceedings of the British Academy, **110**, 93–117, © The British Academy 2001.

— found in much the same region. The enormous impact of Palaeolithic art upon our view of its creators cannot be doubted. But at the same time the Franco-Cantabrian style has a limited extent in space. And while the simpler forms of rock shelter art may be virtually a worldwide phenomenon during this time, that is certainly not true of images with the sweep and coherence of those from Lascaux or Altamira or from the Grotte Chauvet which so much impress us today. As Mellars (1991: 63) has shown, there are plenty of other significant innovations associated specifically with *Homo sapiens*, including a shift in lithic production from 'flake' to 'blade' technology, the use of carefully shaped bone, antler, and ivory artefacts, the increased tempo of technological change, the greater degree of regional diversification, the appearance of a wide range of personal adornments including beads and pendants (White 1989, 1993), and the development of customs of deliberate burial. Greater mobility is indicated by the greater distances from which raw materials, such as specially selected flint, were obtained.

But against these undoubted innovations, and others that can be indicated in different parts of the world, it remains the case that (apart from the Franco-Cantabrian cave art) the differences are not such as would greatly interest either untutored laymen (among whom I would situate myself, so far as Palaeolithic archaeology is concerned) or the perceptive extra-terrestrial observer casually visiting our planet.

If, on the other hand, one surveys the products of the past six or seven thousand years in different parts of the world, one is impressed by a whole range of notable achievements which evidently place our species in a different class from the rest of the animal kingdom: by temples and pyramids in Egypt, early cities and ziggurats in the Near East, great cities of the Indus, and the complex societies and technologies documented in China already from the Shang period of about 1600 BC. In the Americas we may draw attention to the spectacular accomplishments of the Incas of Peru and the Aztecs of Mexico at the time of the Spanish conquest and to the wide range of products of their predecessors. These seem, at least at first sight, advances of quite a different and more remarkable order.

It would seem then that the arrival of our species over much of the surface of the globe did not produce any very remarkable consequences for several tens of millennia. This then is the paradox. If human societies of the early Upper Palaeolithic period had this new capacity for innovation and creativity which notionally accompanies our species, why do we not hear more about them?

Put so baldly the questions underlying the generalization may be a little oversimplified. But it does seem to be the case that in many parts of the world there is indeed a hiatus accompanying what has sometimes been termed, following Gordon Childe, the 'Neolithic Revolution' (Childe 1936). On closer inspection, however, as Ofer Bar-Yosef effectively demonstrates in his chapter in this volume, many of the key steps in the development of sedentism are seen before the domestication of plants and animals was effected. Moreover among just a few hunter-gatherer communities in more recent days, such as the Native Americans of the north-west of America, an abundant food supply has permitted the development of sedentism and of a more complex social order of the kind which one more readily associates with the life of farming communities. It may be suggested therefore that it is sedentism rather than agriculture which marks the more significant change.

The more complex behaviour which we see rather widely in such circumstances may thus have the presence of our species, *Homo sapiens*, as a necessary condition. But evidently that presence is not a sufficient condition for the development of more complex behaviours.

If, by the notion of 'the human revolution' we do indeed intend an evident, obvious, and significant change in society and in material culture, this is therefore not a feature which can be ascribed simply to the appearance of our species. Many of the concepts which we associate with the notion of 'mind' — the more complex range of behaviours, the use of a wide range of symbols, the development of complex notations such as writing, permitting the emergence of a collective memory and the whole phenomenon of what Merlin Donald (1991) has termed 'External Symbolic Storage' — are later developments.

This leads me to suggest that we should regard this supposed human revolution and probably the emergence of 'mind' itself, as a process which, while it may have begun (at least in some respects) with the emergence of our species, has in fact to be regarded as a more gradual one, operating in several phases and stages, and perhaps independently in different parts of the world. For does it make sense to speak of the full development of 'mind' if we are not yet in the presence of complex notations, and the sort of argumentation, for instance in the field of mathematics and astronomy, which only writing permits?

I would further say that it would be a mistake to over-privilege writing in these matters. Between the origins of sedentism and of writing there were at least 5,000 years of development in which material culture was used for a number of symbolic purposes in western Asia, and a comparable span of time (although later in calendar years) separates the onset of sedentism in Europe from the inception of literacy. Indeed, one of the purposes of this chapter is to draw attention to the early developments in the use of material culture in developing concepts of prestige and commodity, and to the social interactions and institutions which accompany the construction of monuments in pre-literate societies.

The assertion that 'mind' is a feature which, in the broad span of human history, develops more fully only with the onset of sedentism should not be taken as a disparagement of the status of recent and contemporary mobile hunter-gatherer groups. In the first place, these have as long an evolutionary history as do more complex societies. Fifty centuries are fifty centuries whether

among mobile or sedentary societies. But the crucial reality is that the onto-
genesis of mind within our own society is relived again as every child learns
to see and understand, to speak and to learn more complex concepts and
behaviours. As Edwin Muir put it:

> Yet still from Eden springs the root
> As clean as on the starting day.

Every normal child has these potential capacities, although among the under-
privileged not all are fulfilled. I realize that the proposition that 'mind' is in
some senses less fully developed among the illiterate and innumerate in our own
time is a potentially controversial one, open to misinterpretation. But it does
seem a conclusion which in a sense is the consequence of the view that the
'hardware' of the entire human species has changed little in the past 40,000
years. In that sense we are all, and were all, born equal. What varies is the 'software', the learned patterns of thought and behaviour whose nature depends
crucially upon the society and the specific circumstances into which we are
born.

The true human revolution came only much later than the emergence of the
species, with the development of a way of life which permitted a much greater
engagement between the human animal and the world in which we live. Human
culture became more substantive, more material. We came to use the world in
new ways, and became involved with it in new ways. The trigger for this new
embodiment, this new materialization, may have been sedentism.

ENGAGEMENT AND INSTITUTIONAL FACTS

The engagement of which we speak implies the development of new interrela-
tionships between humans and the material world. Most animal species may be
thought of as browsers and collectors, dependent mainly upon plant food, or
as hunters who in many cases need to catch the highly mobile prey upon which
they depend. The same is true of most hunter-gatherers, although they have
indeed their own culturally mediated forms of engagement. The development
of stone and wooden tools and of the important device of fire already in earlier
phases of hominid existence are important early steps in the engagement
process. The development of new hunting strategies and of new tool kits in the
Upper Palaeolithic are further such steps. The use of the bow and arrow, utiliz-
ing the elastic properties of the string of the bow in order to make a more effective projectile, is a beautiful example. The increasing distances over which raw
materials were sought is another feature of the process of increasing engagement. Nor should one overlook the efficacity of social developments, such as
the use of larger and more specialized hunting parties to catch and kill big

game. These no doubt depended upon a number of technical advances, but it was the skill and effectiveness in communication involved which allowed the more productive functioning of the social unit, without which the hunting strategy would not have worked.

I shall argue, however, that it was not until the development of sedentism that a much wider range of processes involving new kinds of engagement came into play. The exploitation of domestic plants and animals is clearly prominent among these. But so is the development of new technologies beyond the novel biotechnologies of domestication. The most obvious of these are the pyrotechnologies. Already before the inception of food production we see occasional instances of the use of fire to modify raw materials, to produce pottery and baked clay figurines. Significantly, with the Jomon pottery of Japan, just as in the terracotta figures of Gravettian Pavlov and Dolni Vestonice, we are speaking of what may have been partly sedentary communities. It was from the skills of the potter that those of the smith are likely to have emerged. All of these represent processes of more elaborate and developed engagement.

It would be a mistake, however, to exaggerate the technological dimension without taking sufficient note of the fact that nearly every such technological innovation is also a social one. It is its use as much as the technique of production which characterizes a new innovation, as the history of metallurgy clearly shows. It is not uncommon for technological advances of great potential value to lie unexploited for centuries. The celebrated example of the wheeled toys in Mesoamerica is a case in point. The wheel was not used in the Americas until the time of the European invasions. (But there of course there may have been limitations of traction. The wheel was not much used in the Old World either unless accompanied by the ox or the horse.)

The key point, however, is that the social context, the necessary matrix for the development of technological innovations during the increasing engagement with the material world, is dependent upon social relationships which in many cases are based upon cognitive advances. They depend upon values, ordered values, and upon rules of conduct and behaviour. These in turn are regulated by social roles and by distinctions of status. Many of these social realities depend upon what may be termed 'institutional facts'. This is an important nub in the argument. For when analysed in detail, most new forms of engagement between humans and the material world prove to involve also a cognitive basis. They are dependent upon shared understandings among humans within a community, understandings which are at once social and cognitive. They depend in many cases upon the use of symbols. Many of these are abstract concepts which may readily be given verbal expression. Marriage could be one of these; property, debt, obligation would be others. And, although in a sense abstract, most of these also have a very real and physical reality. As we shall see, there are some cases where there is an inherent link between the physical and material and the symbolic. The concept of weight would be one such. Units of weight are indeed conceptual, but they would be unthinkable without the experience of the physical reality of weight, and the experience that 'more heavy' and 'less heavy' are repeatable observations which may be compared, balanced and quantified. I would like provisionally to suggest the notion of *constitutive* symbol (Renfrew 2001) where the symbolic or cognitive element and the material element co-exist, are in a sense immanent, and where the one does not make sense without the other.

The philosopher John Searle (1995: 31ff) in *The Construction of Social Reality* has drawn attention to the key role of what he terms *institutional facts* which are realities by which society is governed. As he puts it (Searle 1995: 27)

> Some rules regulate antecedently existing activities . . . However some rules do not merely regulate: they also create the very possibility of certain activities. Thus the rules of chess do not regulate an antecedently existing activity . . . Rather the rules of chess create the very possibility of playing chess. The rules are *constitutive* of chess in the sense that playing chess is constituted in part by acting in accord with the rules.

The institutional facts to which Searle refers and which are the building blocks of society include such social realities as marriage, kinship, property value, law, and so forth. Most of these are concepts which are formulated in words and which are best expressed by words — that is how Searle sees it. He draws attention to what he terms the self-referentiality of many social concepts, and he takes 'money' as a prime example. But the point which I want to stress is that in some cases — and money is a very good example — the material reality, the material symbol, takes precedence. The concept is meaningless without the actual substance (or at least it was in the case of money for many centuries, until further systems of rules allowed promissory notes to become formalized as paper money, then as equities and bank cheques, and now as electronic transactions). In early society you could not have money unless you had valuables to serve as money (Powell 1996), and the valuables (the material) preceded the concept (money).

Institutional facts as material reality

Some material symbols, then, are constitutive in their material reality. They are not disembodied verbal concepts, or not initially. They have an indissoluble reality of substance. They are substantive. The symbol (in its real, actual substance) actually precedes the concept. Or, if that is almost claiming too much, they are self-referential. The symbol cannot exist without the substance, and the material reality of the substance precedes the symbolic role which is ascribed to it when it comes to embody such an institutional fact.

Most workers privilege the functioning of language in the emergence (i.e. coming into being) of institutional facts, and Searle himself, as a philosopher, is perhaps preoccupied with the operation of words. The same criticism may be levelled against Merlin Donald (1991), whose useful concept of 'External Symbolic Storage' is too readily equated with the use of writing (Renfrew: 1998), although it cannot of course be doubted that writing is indeed ultimately the most efficient form of External Symbolic Storage, at any rate until electronic means became available. Indeed the role of artefacts as central players in the story of human symbolic evolution is often undervalued (e.g. Lock & Peters: 1999). However, it is the case that in many instances it is the engagement process itself (between cognizant human individuals and groups, and material culture) which brings about the emergence of a new cognitive dimension. As noted above and further discussed below the very notion of 'weight' does not emerge as a word or a pre-existing concept imbued with 'meaning'. Weight is not conceivable without the experience of heavy matter. The notion of equivalent lumps of matter which may be equated in terms of some inherent property (which we term 'weight' or 'mass') comes about through sentient and cognizant human experience. The experience is not preceded by the concept, and for that reason I find the term 'materialization' (De Marrais et al. 1996), which might be regarded as the process of translating concept into matter, less satisfactory than 'engagement'. There are many features of the real, material world which can easily lead, in interaction with the human mind, to the development of new relationships of engagement which involve a conceptual as well as a material dimension. 'Value' is another such: it is difficult to conceive of value without first having some experience of valuables — that is to say of things to which value may be ascribed. It may similarly be argued that the human capacity for categorization is in part the product of experience of the natural world, where plant and animal species present the obvious lesson that living things present themselves already in what are effectively categories.

To some extent to make such an observation may be described as taking a phenomenological approach. But, if so, it is a phenomenological approach which is concerned primarily with the time dimension. It is one which seeks to understand why some societies develop such concepts and why others do not. That is very much a social question, and is not simply a matter of the human individual as a timeless being standing alone in the face of the universe.

It should be noted that this approach, which emphasizes the material reality of many institutional facts and the relationship between the conceptual and the material often involved in the process of engagement, has the potentiality of transcending the traditional mind/matter dualism, which remains very much a feature of contemporary archaeological thinking. For the early New Archaeology shared with traditional Marxist approaches a preoccupation with subsistence and production. It developed a standpoint which might reasonably be termed 'materialist'. In that sense it was functionalist and developed a position which has been described as 'functional-processual' (Renfrew & Bahn 2000: 491–5). The so-called 'post-processual' critique of the New Archaeology made precisely that observation, and developed an alternative 'interpretive' approach, in which the key concept and desideratum was 'meaning'. In many ways this 'interpretive' approach, which could certainly be described as 'idealist' in the sense which Marx attacked in *The German Ideology* (Marx & Engels 1977), has at times failed to follow the insights into the human interactions with the material world which were analysed in the functional-processual tradition. It should be noted that, when appropriately applied, the approach advocated here privileges neither the materialist nor the idealist, neither matter nor mind. It works within a cognitive framework but not simply a mentalist one. In the course of the engagement process new relationships between humans and the material world emerge. They are at the same time social relationships and therefore operate between human and human, and they are also cognitive.

The process is sometimes a progressive one, for within it matter comes to be seen as possessing new properties. Indeed among the first of these is the notion of 'property' itself (which is further discussed later in this chapter). What at first sight seems a trivial play on words — that things have or suggest properties (i.e. aspects or features), amongst which is the capacity to be owned (to become 'property') — is more interesting than that. For this is a duality of meaning shared also in Greek: *ousia* has both senses, and this is so in other languages also (cf. German *Eigenschaft*/*Eigentum*). These properties can be at once material and conceptual. And part of the process of the human exploration of the material world is indeed the discovery of new properties which permit the development not only of new technologies (ceramics, weaving, metallurgy, electrical engineering, transport, radio) but of new social relations also.

It is my argument that this process of engagement lies at the nub of the development of human societies. Moreover in non-literate societies it is material symbols which play a central role by allowing the emergence and development of institutional facts. Some classes of institutional fact may well be a feature of all human societies including hunter-gatherer societies. Affinal kinship relations — including the institution of marriage or of something like it — seem to be a feature of all human societies. But I shall argue that many kinds of material symbol are not generally a feature of mobile hunter-gatherer societies. It is not until the emergence of sedentary societies (usually in conjunction with food production) that the process of the human engagement with the material world takes on a new form and permits the development of new modes of interaction with the material world, allowing the ascription of (symbolic) meaning to material objects.

This, I would argue, is the solution of the Sapient Paradox — why so little that was truly and radically novel initially accompanied the emergence of our own species *Homo sapiens*, despite what we can now recognize as its enormous inherent potential to undergo and initiate radical change.

MATERIAL AND COGNITIVE CONSEQUENCES OF SEDENTISM

What seems a simple shift, from the mobile life pattern of most hunter-gatherer communities to one of sedentism, is in reality one with very significant consequences. Sedentism implies, of course, living in one place on a permanent basis — or at least for several years at a time. It therefore implies a permanent place of residence. Usually that place will be a house — a deliberate residential construction, requiring input of both labour and materials. The way is open now for the development of permanent installations — storage facilities, preparation facilities requiring heavy equipment, locations (such as ovens) for the application of special techniques, and so forth. The way is open also for the storage of property, and hence for the emergence of commodities.

Of course there exist partially mobile economies, for instance those relying upon transhumance, where some of these things are possible. And there are other adaptations, such as those of nomadic pastoralists, which show some of the features of sedentary societies.

Most obviously, sedentism requires the availability of a mix of food resources permitting year-round occupancy. In most cases this implies food production (although as noted above marine and aquatic resources can sometimes support sedentism without food production). It also implies storage, for instance of hazelnuts in Mesolithic northern Europe and of cereals in the Near East.

Sedentism favours the development of 'property'. The stored foodstuffs are critical to survival. The house constructed by one group continues to be occupied by that group which has preferential access to it. The domestic animals reared by one group will usually be theirs to exploit and slaughter — their property. Access to the land cultivated by the group and to its products may well be restricted — who sows may reap. It is easy to see how the 'institutional fact' of 'property' emerges through a substantive material reality before it becomes a legal concept. Property is one of those special concepts discussed further below which are at once symbolic and material and may be described as constitutive symbols.

Ian Hodder in *The Domestication of Europe* (Hodder 1990) has emphasized very effectively the profound change in lifestyle that accompanies the spread of the *domus*, the home of a sedentary population. But as already noted, while food production is a concomitant of much sedentary life, it is not so much food production as sedentism on a stable and enduring basis which is the revolutionary component of the 'Neolithic Revolution'.

The process of engagement or substantivization continues with the development of the new technologies involved. The use of heavy stone querns for grinding is difficult in a mobile economy (since the querns are too heavy to be transported). The use of dried mud (*tauf* or *pisé*) becomes feasible as a building material, laying open the possibility of large constructional complexes such as that seen at Çatalhöyük. Extensive stone construction is no longer unduly labour-intensive, if it is to be used over a long time period. Not only do such factors make possible, but the scale of investment makes desirable, the development of defensive facilities, such as the very early walls at Jericho.

Sedentism is also associated with what Jacques Cauvin has termed 'the birth of the divinities' (Cauvin 1987, 1994), and it is to be noted that these occur in the Near East in settlements prior to the development of domesticated plants and animals. Indeed Cauvin has suggested that there may be a causal relationship. It should be noted that to be altogether effective these divinities need to take material form — this is the 'materialization' process noted by De Marrais *et al.* (1996). A related point is made by Mithen (1998) in relation to the long-term persistence of religious beliefs, which is facilitated by their permanent embodiment in material form.

The reference above to the use of installations in sedentary societies leads on to what was one of the most significant of these, the oven. The oven represents a new development in pyrotechnology, which was already significant in hunter-gatherer communities for cooking, for the heat pre-treatment of flint, and in other ways. But while the oven may have been an extension of the open fire in the field of food preparation and cooking (for parching grain and baking bread), these new enterprises led on to the development of new materials. Pottery manufacture is seen in most sedentary societies but few mobile ones, ceramic containers being too heavy and breakable for transportation, while those of string, bark, wood, or leather were more practical. And in Europe as in the Near East it is clear that the pyrotechnology required for ceramic production soon offered the technical means needed for metallurgy. With the development of ceramics and of metallurgy came the production of the first artificially produced materials. In the case of copper and gold (and later of silver, and of bronze) these led on to the crucial nexus surrounding prestige goods — value, measure, and exchange — as further discussed below.

GROUP-ORIENTED AND INDIVIDUALIZING SOCIETIES

In early food-producing societies, and indeed in more complex prehistoric societies, it is possible to make a rather basic distinction which, while not of universal validity, does reflect a difference which is widely seen (Renfrew 1974) and which has conveniently obvious archaeological correlates.

Some early societies appear to assign very little personal importance to prominent individuals. There is no evidence of salient ranking. On the contrary, so far as personal equipment and adornment go, they might at first sight be described by the anthropologist as 'egalitarian' societies. But at the same time such societies perhaps had what must have been a pronounced social structure. They are often more than simple dispersed farmsteads without any overarching social articulation. Instead they are in some cases capable of significant collective action. In practice such group action is often most evident to the archaeologist in the form of collective works. As Edmund Leach (1954) has indicated, in traditional Burma there were irrigation projects which required collective endeavour on a considerable scale, far exceeding the resources of the single farmstead or even the single village. In the prehistoric record of north-western Europe there are substantial stone structures, frequently termed 'megalithic', whose construction required significant group endeavour. The chambered cairns of north-western Europe, dating to the Neolithic period, at the more modest end of the scale, must have required a labour input of some 10,000 work hours. The larger henge monuments of southern Britain may have needed as many as one million work hours. And it has been calculated that the largest monuments of the time, such as Silbury Hill and Stonehenge, would have needed tens of millions of work hours when the transportation of raw materials as well as the construction is taken into account (Renfrew 1973: 548).

Yet these societies in general do not give us very much trace of the individuals involved. These were certainly not state societies. They are not accompanied by rich burials nor by any kind of finery. Prestige goods, such as polished stone axes of attractive materials, are not in general found associated with burials. Whether or not it is appropriate to designate societies whose achievements imply considerable managerial resources as 'chiefdoms' is a matter for discussion. Certainly one does not see any archaeological record of the presence of the chief in person. But the group achievement is evident. For that reason the term 'group-oriented' is appropriate. Among examples to be quoted are the henge monuments of southern Britain, including Avebury and Stonehenge. At the northern extreme there is the impressive complex on Orkney which includes the Ring of Brodgar and the Stones of Stenness (both henge monuments) and the impressive passage grave of Maes Howe. The so-called 'Temples' of prehistoric Malta would be a further case in point. But the observation holds more widely. In the American south-west the great structures of

Chaco Canyon, dating from the early second millennium AD are the evident result of concerted group activity. But, with a few exceptions, they betray very little sign of prominent individuals of high status. That there was a managerial capacity no one can doubt, but it was not centred upon the person of an indiviudal who was accorded prominent high status, celebrated by conspicuous symbolic artefacts.

At the other end of the scale we do find, in many early societies of a rather different character, that there were prominent individuals whose high status was celebrated by the possession of a finery of rich artefacts made of exotic materials and fashioned into shapes of evidently symbolic significance. These we may term individualizing societies. In the Early Bronze Age of north-west Europe, to begin with a period succeeding that of the megalithic monuments discussed above, we find individual burials under round mounds accompanied by bronze weapons indicative of high status, and sometimes accompanied also by gold ornaments reinforcing that impression. Even earlier, in the Copper Age cemetery of Varna in Bulgaria, where gold makes its first major appearance in human history, there are burials which today seem dazzling in their wealth. The shaft graves at Mycenae, at a rather later date, give a comparable impression that the high-status individuals buried there were keen to enhance their personal prestige by processes of conspicuous display and consumption. In the New World there are many cases of the conspicuous burial of high-status individuals. The civilizations of Mesoamerica give many examples, of which the burial of Pacal at Palenque is perhaps the most celebrated. Of course we are dealing here with a very complex society, a state society. It distinguished the importance of such high-prestige individuals not only by rich burial but by monuments of considerable grandeur, and indeed in the Maya case by the erection of stelae bearing inscriptions celebrating significant events in the lives of these rulers.

There is the risk, when the archaeologist discusses what one may term 'individualizing' societies, that we are inclined to place rather too much reliance upon burial data. Clearly one cannot have grave goods involving high-prestige objects unless there is a burial containing grave goods in the first place. There is the risk therefore that the archaeologist may make generalizations about social status and 'individualizing' tendencies on the basis of data which are, in reality, governed to a considerable degree by customs in burial practice. Aspects of such a criticism are valid. If one is dealing with a society, and there are many such, where the remains of the deceased are not ultimately placed below ground, then they are not likely to be recovered. On the other hand the nature of the burial custom is not itself an independent variable, and the development of individual burials, with or without prestige goods, does in itself imply an outlook where it is appropriate to distinguish the individual from the group. So while this factor should be borne in mind, it does not in itself invalidate the argument.

THE PRESTIGE NEXUS: VALUE AND COMMODITY IN
INDIVIDUALIZING SOCIETIES

It is a feature of hunter-gatherer societies that, while there were certainly materials of value, such as workable flint, whose procurement was worth a good deal of effort, the expression of personal prestige through exotic materials was limited in its range. Certainly marine shells were prized and were traded over considerable distances, and ornaments and pendants using them are found, sometimes in burials (White 1989, 1993). The individuality of the person thus found expression by this means. If we are talking of 'individualizing' through the use of material culture, this certainly began in the Palaeolithic period. It is, however, fair to say that in Europe and in western Asia it is not until the early development of metallurgy that we find a range of burials with accumulations of grave goods in a diversity of materials which could in that sense be considered 'rich'. Susan Shennan (1975; see also Lesure 1999) has analaysed cemeteries of the Neolithic period as well as of the Early Bronze Age, and differentiation among the graves is evident already in the Neolithic. But there are few if any cases where one could speak of burials of high prestige or conspicuous wealth.

Such features make their appearance, so far as European prehistory is concerned, in the late Neolithic (or 'Chalcolithic') cemetery of Varna in Bulgaria in the fifth millennium BC. There are graves which have a range of impressive grave goods, even before the objects of copper and gold are taken into account (Renfrew 1986). These include quantities of marine shell and exceptionally long blades of flint which must have been the product of very considerable craft skill. However, it is the quantities of gold found at Varna which bring it first to more general attention. It is notable also that it is in this context that we see a range of copper artefacts. Their use seems here to be as indicators of high prestige. It was perhaps only later that copper became a really useful material, and not until its alloying with tin that it was significantly more useful than stone. In a recent paper (Renfrew 2001) I have drawn attention to the interrelationship of four concepts which may have emerged together at that time, although some of them will have had earlier antecedents (see Figure 1).

The Varna cemetery shows very clearly the emergence of a new material which henceforth in Europe would be considered to be of high value: gold. Its ownership and conspicuous display, for instance in the form of such artefacts as are seen at Varna, reflect and confer prestige. Of course it has to be demonstrated that gold was indeed valued highly at the time in question: there is no need to assume that just because we value it highly there was a similar evaluation six thousand years ago. However there are plausible arguments (Renfrew 1986) for arguing this without making any *a priori* assumptions. In Europe there is an interesting link between objects of high prestige, including such new

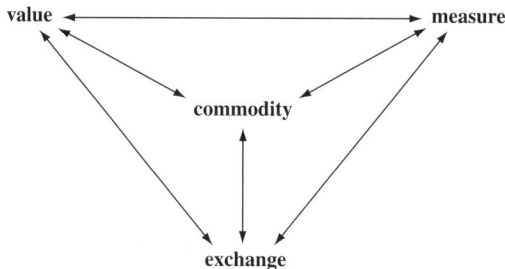

Figure 1. The crucial nexus surrounding the concept of commodity.

weapons of war as daggers and swords, and the development of the new metal-lurgical industry. So obvious does it appear to us today that gold is valuable that we speak of its 'intrinsic' value. But in reality all values are ascribed — they are indeed institutional facts for the communities in question. At the suggestion of Arjun Appadurai (Renfrew 1986) it is perhaps preferable to speak of 'prime value' rather than of intrinsic value. The association of value and prestige becomes a crucial one in many societies (e.g. Voutsaki 1997).

As noted earlier, the very notion of weight is a conceptual formulation which arises both from the properties of the material world in which we live and from the human initiative in devising cognitive categories for that world. A metal or stone weight, such as are found in the Indus Valley civilization (Renfrew 1982), has a symbolic role as part of a measurement system used for imposing order upon the world. But its origin is not in some abstract verbal formulation: it comes from the properties of the world itself. This is a case of what was earlier termed a constitutive symbol.

There is little point in having a process of 'weighing', or in creating 'weights', unless there is something to be weighed. In many cases the purpose of weighing is to establish an equivalence between different materials in terms of this measurable parameter, weight. In such instances the equivalence is part of a conceptual structure where an exchange between two materials is organized in such a way that quantity X of material A is regarded as of equivalent value with quantity Y of material B. We are on the brink here of an exchange system which goes beyond unique barter events where bundle P is agreed to be worth bartering for bundle Q, without any more specific analysis or measure of the content of the bundles. However, it is not difficult for us to see the functional relationships between exchange, value, and measure (in this case weight). It is in such structured systems of exchange that different materials lose their uniqueness — in the sense of this particular piece of gold, or that particular bag of grain — and become commodities. We have reached the point where a particular amount of gold (any good-quality gold, not this particular piece of gold) may be equated with that much good-quality grain (defined by quantity rather than as a specific and unique bundle). Gold and grain have now become commodities. In many complex societies the emergence of commodities, of raw materials widely traded and exchanged, is a significant development (e.g. Gregory 1982; Sherratt & Sherratt 1991).

This is a crucial nexus for the development of any economic system (see also Michailidou 1999). And it is a good example of the sort of engagement process discussed earlier, where concepts formulated by humans in the light of their experience of the world are used to modulate the way those humans deal with the material world. The notion of commodity — that we can speak of wheat in the general sense rather than as a series of unique bundles of the material — is a conceptual advance. The way the society chooses to deal with that — by weight, or by solid volume measure — and the units used, are specific institutional facts. The notion of cumulative value — that ten kilos of wheat are worth ten times more than one kilo of wheat — although intuitively obvious to us, is again an institutional fact, albeit one that is based on underlying general mathematical concepts. And of course the concept of exchange, the notion that in a well-ordered world quantity X of material A is worth quantity Y of material B, is again an institutional fact.

These are the basic underlying structural features of human societies. Some of them may be near-universals. But many of them are no doubt specific to specific societies, or rather to regionally specific trajectories of development, since such conventions, such institutional facts, show great temporal stability. Nor need they be prosaic and material, as in the example given. In Mesoamerica, in different civilizations, there is considerable conformity among views of how the cosmos is ordered, about the nature of the four quarters, that there is an afterworld which is an underworld and so forth. If these became enduring realities, albeit cognitive realities, for the societies in question, then they were institutional facts. In our own society there are sixty seconds in a minute and sixty minutes in an hour, yet few of us bother to remember that these were arbitrary decisions made by astronomers in ancient Babylon. They are among the institutional facts of our own society. This last point may be dependent upon notational systems, but the other examples given here are not in any way dependent upon writing. All of them emerged, in many different trajectories of development, in prehistoric times.

These then are instances of the way human development comes about through increasing engagement with the material world, mediated by institutional facts. And I would go so far as to claim that the development of such notions as measurement (and of units of measure) and as equivalence in an exchange transaction are important components of 'mind', seen as something which develops with the human story rather than emerging full-grown with the formation of our species.

THE ACTIVE SOCIAL ROLE OF MONUMENTS

Among the group-oriented prehistoric societies of north-western Europe, as noted earlier, stone monuments play a conspicuous role. They vary in scale from the earthen long barrows of southern England and the stone chambered cairns of Scotland to the very much larger henge monuments, some of which, like Stonehenge and Avebury, contain circles of standing stones. In earlier generations these were seen as the result of the migrations of peoples or the diffusion of ideas from more civilized lands. Now, on the contrary, they are seen as local products. One view of the long barrows and chambered cairns, which evidently served as tombs, is that they were 'territorial markers of segmentary societies'. Such a view uses the apparent regularities in their spatial distribution to suggest that each is associated with the habitual territory of a resident population. The notion of 'segmentary' society implies little more than that these were small, autonomous social units of comparable size to their neighbours. Often the larger monuments have been seen in similar terms, reflecting the growth of larger social units in the later Neolithic period, while the chambered cairns date back to the earlier Neolithic.

Such a view does not seriously conflict with the available evidence. But it might none the less be criticized as somewhat 'reflective', in the sense that it interprets the monuments as *reflecting* the existing social structure. Segmentary societies, it is argued, often need a ritual and ceremonial focus, and this need was met by these local centres. In the same way, group-oriented 'chiefdoms' (if that term is felt appropriate) need a centre, and the great henges served as ceremonial centres and perhaps as pilgrimage centres also for their parent communities. Thus they too would reflect aspects of the social order. It was, however, Ian Hodder who many years ago (Hodder 1986) emphasized the active role of material culture. Culture is not seen as something which merely reflects the social reality: it is part of the process by which that reality is constituted. The general position adopted here is very much in harmony with such a view. The development of social institutions is here seen as part of the process of the increasing engagement of humans with the material world. It is in the course of this engagement that new institutional facts are called into being, and new social institutions initiated.

We can apply this line of thought to Neolithic Britain. In the case of the chambered cairns and long barrows we can suggest that, rather than reflecting a pre-existing social order, they helped to call that order into being. At the time of its first inception the long barrow or the cairn will have initiated a project, and one which would in due course involve some 10,000 work hours. In order to bring this about, the occupants of the territory in question would need to invest a great deal of their time. They might need also to invoke the aid of neighbours in adjoining territories, encouraged no doubt by feasting and a local celebration. One may imagine that when the monument was completed it might itself have been the locus for further, annual celebrations and feast days. It served as a burying place and as a social focus for the territory. The suggestion here is that it was as a result of these ongoing social activities, along with other activities of a ritual or religious nature, that the cairn or barrow came to be the centre of what soon emerged as a community. Yet it is reasonable to suggest that this community would not have come into being had it not been for the ongoing activities centred upon the cairn. Such a view is not far from the 'structuration' approach advocated by Barrett (1994). In the present context it suggests how a particular form of engagement with the material world — the construction and varied use of a burial cairn — could help promote the emergence of a coherent new social unit. The same point applies with even greater weight, on a larger scale, where the henge monuments are concerned. Their construction certainly implies some pooling together of labour of a number of the smaller, earlier territories. But once the henge was built it could serve as a focal point for those territories. This too would be an example of the active role of material culture. It would reflect too a new kind of engagement, where a larger group of people would use this constructed monument for ritual, social, and perhaps religious purposes. The end product could be the emergence of a coherent larger community where none was before.

In considering the possible emergence of group-oriented societies in this way, centring upon the construction of a regional or territorial monument, it is worth asking further about what precisely it is that is so attractive about a circle of stones that it should act as the centre for important rituals (as we are suggesting) and eventually become the central focus for an emergent new social unit.

One answer must come from the affective power of a monumental construction to impress us with its material presence. Such a construction, as a number of authors have recently stressed (e.g. Bradley 1990), brings with it, or rather markedly enhances, a sense of place. It is a tenet of much current archaeological thinking (e.g. Tilley 1994) that the landscape in which we live and work is not a given; it is a constructed environment, rich with the memories of earlier people and earlier events. Even without man-made constructions, the accretion of these spatially specific memories makes the landscape as much a social as a physical reality. The insertion into this landscape of the memory of a monumental construction reinforces that process. It might be an exaggeration to suggest that the emergence into nationhood of the state now called Zimbabwe (formerly Southern Rhodesia) was a product of the earlier construction of the monument known as Great Zimbabwe. Yet at the same time the achievements of the indigenous ancestors of the area will certainly have played some role in the self-recognition, the ethnicity, of the population concerned.

There is in the construction of even the simplest of monuments, as the recent work of the sculptor Richard Long has shown (Renfrew 1997: 10), something which attracts our interest and engages our emotions. This too is a kind of engagement with the natural and material world. It is an action which is more symbolic than practical. And again one may think of constitutive or immanent symbolism, for it is not initially clear just what the constructed feature symbolizes. It just is. And it serves as a marker for the actions of its maker and of what that maker wished to remember — which is precisely what a monument is. Later it can take on a more explicit meaning, serving to represent and indeed 'symbolize' the community whose emergence it has facilitated.

Here I am reminded of that rather mystical quality of material things, in some cases specific material things, which Marx (1886) discussed when he wrote of 'the fetishism of commodities'. This could well have been spoken of in the previous section when discussing the value and prestige inherent in artefacts made of precious materials. But it is interesting, and I hope not too frivolous, to think of an analogous, almost mystical quality possessed by monuments.

In the passage which follows, taken directly from *Das Kapital,* I have substituted the word 'monument', highlighted in italics, for the word 'commodity' as employed by Marx (1886: 76–7):

THE FETISHISM OF *MONUMENTS* AND THE SECRET THEREOF

Whence, then, arises the enigmatical character of the product of labour, so soon as it assumes *monumental* form? Clearly from this form itself. The equality of all sorts of human labour is expressed objectively by their products all being equally valued; the measure of the expenditure of labour-power by the duration of that expenditure, takes the form of the quantity of value of the products of labour; and finally, the mutual relations of the producers, within which the social character of their labour affirms itself, take the form of a social relation between the products.

A *monument* is therefore a mysterious thing, simply because in it the social character of men's labour appears to them as an objective character stamped upon the product of that labour; because the relation of the producers to the sum total of their own labour is presented to them as a social relation, existing not between themselves, but between the products of their labour. This is the reason why the products of labour become *monuments,* social things whose qualities are at the same time perceptible and imperceptible by the senses. In the same way the light from an object is perceived by us not as the subjective excitation of our optic nerve, but as the objective form of something outside the eye itself . . .

There is a definite social relation between men, that assumes, in their eyes, the fantastic form of a relation between things. In order, therefore, to find an analogy, we must have recourse to the mist-enveloped regions of the religious world. In that world the productions of the human brain appear as independent beings endowed with life, and entering into relation both with one another and the human race. So it is in the world of *monuments* with the products of men's hands. This I shall call Fetishism which attracts itself to the products of labour, so soon as they are produced as *monuments*, and which is therefore inseparable from the production of *monuments*.

This Fetishism of *monuments* has its origin, as the foregoing analysis has already shown, in the peculiar social character of the labour that produces them.

I find this juxtaposition suggestive. Marx, of course, emphasizes labour in his discussion. But if we broaden that concept very slightly to think instead of human endeavour, we see then what, in the context of this chapter, I would suggest we might read as a meditation upon the power of constitutive symbols. In the case of a monument like Stonehenge or the Ring of Brodgar, it is the concentration of human endeavour and labour which finds expressive form in a massive artefact of material culture. This is a highly conspicuous instance of the engagement of humans with the material world. The visitor even today finds it prodigiously impressive. How much more so must it have been when it commemorated people whose histories were recorded by oral tradition, and when those viewing it felt it to be the historic centre of the community to which they themselves still belonged.

Although this discussion is a very incomplete one, I feel that this passage of Marx, which has long been recognized as a seminal one in relation to material goods (commodities), does offer insight when transposed in this somewhat Procrustean way to apply to monuments. It is not too much to say that the role of monuments in group-oriented societies is to some extent replaced, or at least superseded, in individualizing societies by that of high-prestige goods. Of course monuments continued to be built, but in individualizing societies monuments tend to be explicitly directed towards the glory of the ruling individual and they work alongside the princely goods which are now part of the accoutrement of that individual. Thus the pyramids of Egypt worked in the way I have described and in the way that the restructuring of Marx indicates, but they worked now in a society where the individual was at the centre, even if that individual himself, as in the case of the Pharaoh, also represented the social group: *L'état, c'est moi!*

THE EMERGENCE OF PROPERTY

A critical knowledge of the evolution of the idea of property would embody in some respects, the most remarkable portion of the mental history of mankind. (Morgan 1877: 6)

I should like to conclude this paper with a preliminary treatment of the notion of one of the most obvious of institutional facts, and one which relates closely to the engagement between humans and the material world: property. It is one which was considered in a pioneering treatment by Lewis Henry Morgan, to the extent that the observation at the head of this section anticipates a number of the points which I have been seeking to make in this paper.

It seems strange that 'property' as a feature of society is today rarely given much consideration in the archaeological literature. Even in Britain, where there has been so much analysis of the Neolithic and Bronze Age periods, the word is not in frequent usage.

Neither 'property' nor 'ownership' is to be found in the indexes of such influential and thoughtful works as *The Domestication of Europe* (Hodder 1990), *Rethinking the Neolithic* (Thomas 1991), *Altering the Earth* (Bradley 1993), *Fragments from Antiquity: an Archaeology of Social Life in Britain 2900–1200 BC* (Barrett 1994), *Social Being and Time* (Gosden 1994), *A Phenomenology of Landscape* (Tilley 1994), *Europe in the Neolithic: the Creation of New Worlds* (Whittle 1996) or *Archaeological Theory: an Introduction* (Johnson 1999), while 'property transfer' has a single mention, inspired by Jack Goody, in *Economy and Society in Prehistoric Europe* (Sherratt 1997: 195–6). Nor does 'land tenure' or 'tenure' fare very much better.

This, I suspect, is simply because the notion of ownership has been insufficiently problematized. It is to be seen as an 'institutional fact' — it depends upon the customs and beliefs of the community. It is in some senses conceptual, and archaeologists have traditionally been nervous about ascending the Hawkesian ladder up from technology and subsistence to social aspects and then as far as the conceptual rung.

The broader conceptual significance of the role of material objects in Neolithic Britain has certainly not been neglected. Thomas has written of 'an economy of substances', where 'the circulation of people and artifacts from place to place allowed significations that had been established in one context to be transferred to another' (Thomas 1999: 81). He emphasizes the notion of context-specificity and argues that 'the term *economy of substances* should not be taken to infer an overarching structure, so much as a system of signification that might allow the production of meaning in a specific and localized context' (ibid.). Indeed his position in some ways anticipates that taken here: 'The core of the argument will be found in the observation that artefacts are not a mere reflection or product of society, but are integral to social relationships' (Thomas 1996: 141). Bradley & Edmonds (1993) have discussed production and exchange in an original study of the axe trade in Neolithic Britain and Bradley himself has broken new ground with his illuminating analysis of Bronze Age hoards and votive deposits (Bradley 1990). At times, however, the very emphasis upon context-specificity can militate against a diachronic approach, which has inevitably also to be to some extent a comparative approach.

This apparent indifference to the concept of property may, however, also be an enduring reaction against what has long been seen, at least among western archaeologists, as the over-simplifying generalizations of Lewis Henry Morgan (1877) and of Friedrich Engels (1884), both of whom laid considerable emphasis upon the notion of property. Morgan's work was divided into four parts (of which the first remains relevant to any attempt at a 'cognitive' archaeology) in which property played a prominent role:

    I.   Growth of Intelligence through Inventions and Discoveries
    II.   Growth of the Idea of Government
    III. Growth of the Idea of the Family
    IV. Growth of the Idea of Property.

His account was, naturally, based upon a very sketchy knowledge of archaeology, since our discipline was then very new. It relied heavily upon his notion of three 'ethnical' periods — savagery, barbarism, and civilization. These, of course, thinly disguised, reappear in many neo-evolutionary treatments as band (or 'hunter-gatherer'), tribe (or egalitarian society), and state (or stratified society).

Property without domestication

No doubt there is much variety among hunter-gatherer societies in relation to concepts of ownership and property. It is clear that in a mobile community, the individual cannot in general effectively 'own' more than he or she can carry, unless concepts of ownership are developed that allow absenteeism in relation to property.

If one's notion of property is based upon that of enduring and exclusive association, then it should not be overlooked that the first such association is that of pair bonding, which underlies the institutional fact of marriage. Both Morgan and Engels saw that the family patriarch could have a dominant relationship within the family. But Engels, influenced by Bachofen, saw *Das Mutterrecht* as an antecedent stage, in a prescriptive rule which does not make anthropological sense today.

It is sometimes alleged that among some hunter-gatherer communities egalitarian principles militate against an individual being too possessive about any particular artefact. The archaeological record does, however, as noted earlier, document personal adornments from the Upper Palaeolithic period, some of them of a high degree of elaboration. It seems indeed plausible that 'wear' and 'gear' may have constituted the earliest forms of personal property, again arising from enduring association with the individual in question. Wear implies clothing and adornment. Gear implies the kit which one uses on a daily basis and which accompanies one when travelling.

The emphasis here upon domestication (see further below) does suggest for consideration the special case of the dog. The dog is often held to have been the first domesticated animal. What is particularly interesting here is that canine behaviour patterns seem to orient them towards a single dominant human. The relation between human and dog may thus be one of the earliest instances of clear ownership, where one person has an enduring and exclusive association with a special material object.

Property and domestication

It is presumably a fair assumption, when dealing with fields of domesticated crops which have been deliberately sown, or with domestic animals in captivity, that if it is domesticated it must be owned. If we note that this would hardly apply for feral plants or animals, this would otherwise seem a safe generalization. Indeed on this basis the farming revolution must have transformed all previous concepts of ownership, since nearly all the essentials of diet would henceforth have a well-identified owner. Rights of access to wild plants and animals may well be regulated among hunter-gatherer societies, where the notion of territoriality is sometimes well developed, but it would be difficult to discover a methodology to investigate this question archaeologically for early hunter-gatherers.

Sedentism favours ownership. Long association between the human individual or group and a house and a field is one of the most obvious features of ownership, not far removed from the notion that 'possession is nine-tenths of the law'. Another feature is associated with work. There is a natural association between labour and ownership. If I build this house, it is mine. If I make this spear, it is mine. That at once brings us to the practice of agriculture. As noted earlier, the use of arable land in practice involves land tenure, and land tenure inevitably raises questions of property.

In the case of domesticated animals, they involve work. They cannot be allowed to wander freely. To constrain them, to herd or corral them, and indeed to care for them involves in some senses their immediate possession. Once again, unless more sophisticated concepts of property are devised, possession may imply ownership.

These considerations open inviting fields for speculation, and I have the sense that these are areas where much further work remains to be done. Of course, as Fleming (1988) has recognized, the dykes and field boundaries found, for instance, in Britain from the Bronze Age, some of which may go back as far as the Neolithic period, offer the possibility for further consideration of early systems of land tenure.

Recognizing property in the archaeological record

The archaeological record does not often permit the recognition of the human individual, other than in the context of the 'single' burial. There are, however, other rare contexts of the human individual surviving with material goods which one might regard as possessions. The artefacts accompanying the Ice Man found in the Austrian Alps (Fleckinger & Steiner 1998; Spindler 1994) may be considered in that way.

As far as burials are concerned, it is now widely recognized that the goods accompanying the deceased are found in the grave as a result of the actions of the persons or community which buried the body (Bradley 1990: 39). They reflect the burial actions and rituals of the living as much as they pertain to the dead. That being said, however, there is no difficulty for the archaeologist in recognizing what one might describe as a 'rich' or 'prominent' burial. The goods associated with the deceased may not necessarily have been the property of the deceased during life, but the association is none the less there, and it has to be explained effectively in terms of the actions surrounding the burial process. One of the most obvious of these may indeed be the choice of a number of the possessions in the property of the deceased to accompany the corpse at burial. Among these one would certainly expect the 'wear' mentioned above — the dress and adornments — and sometimes also the gear, for instance the weapons of the warrior. It seems likely also that other elements of kit frequently found in the richer burials of European prehistory, for instance the equipment associated with the 'Symposium' accompanying graves of the classical period, may indeed have formed part of the worldly goods habitually used by the deceased individual.

CONCLUSION

This brief aside on property as an institutional fact is incomplete, but it may indicate avenues for further exploration. The central point in relation to this chapter is that property is itself an institutional fact whenever an artefact or a piece of land or indeed anything else has a socially recognized owner. The relationship between ownership and property is a good example of the engagement process between humans and the material world. It operates at the conceptual level: it is a conceptual reality. In order to understand more about the origins of human social institutions we need to analyse further this process of engagement and the working of such institutional facts.

REFERENCES

BARRETT, J.C. 1994: *Fragments from Antiquity: an Archaeology of Social Life in Britain, 2900–1200 BC* (Oxford, Blackwell).
BRADLEY, R. 1990: *The Passage of Arms: an Archaeologist's Analysis of Prehistoric Hoards and Votive Deposits* (Cambridge, Cambridge University Press).
BRADLEY, R. 1993: *Altering the Earth* (Edinburgh, Edinburgh University Press).

BRADLEY, R. & EDMONDS, M. 1993: *Interpreting the Axe Trade: Production and Exchange in Neolithic Britain* (Cambridge, Cambridge University Press).

CAUVIN, J. 1987: L'apparition des divinités. *La Recherche* 194, 1472–80.

CAUVIN, J. 1994: *Naissance des divinités, naissance de l'agriculture* (Paris, CNRS).

CHILDE, V.G. 1936: *Man Makes Himself* (London, Watts).

DE MARRAIS, E., CASTILLO, L.J. & EARLE T. 1996: Ideology, materialization and power strategies. *Current Anthropology* 37 (1), 15–31.

DONALD, M. 1991: *Origins of the Modern Mind* (Cambridge, MA, Harvard University Press).

EDMONDS, M. 1995: *Stone Tools and Society* (London, Batsford).

EDMONDS, M. 1999: *Ancestral Geographies of the Neolithic* (London, Routledge).

ENGELS, F. 1884: *The Origin of the Family, Private Property and the State*, 1961 edn. (Moscow, Foreign Languages Publishing House).

FLECKINGER, A. & STEINER, H. 1998: *The Iceman* (Bolzano, South Tyrol Museum of Archaeology).

FLEMING, A. 1988: *The Dartmoor Reaves* (London, Batsford).

GOSDEN, C. 1994: *Social Being and Time* (Oxford, Blackwell).

GREGORY, C.A.1982: *Gifts and Commodities* (London, Academic Press).

HODDER, I. 1986: *Reading the Past* (Cambridge, Cambridge University Press).

HODDER, I. 1990: *The Domestication of Europe* (Oxford, Blackwell).

JOHNSON, M. 1999: *Archaeological Theory: an Introduction* (Oxford, Blackwell).

LEACH, E.R. 1954: *Political Systems of Highland Burma* (London, Bell).

LESURE, R. 1999: On the genesis of value in early hierarchical societies. In Robb, J.E. (ed.), *Material Symbols, Culture and Economy in Prehistory* (Carbondale, IL, Southern Illinois University), 23–55.

LOCK, A. & PETERS, C.R. (eds.) 1999: *Handbook of Human Symbolic Evolution* (Oxford, Blackwell).

MARX, K. 1886: *Capital*, vol. I, 1974 edn. (London, Lawrence & Wishart).

MARX, K. & ENGELS, F. 1977: *The German Ideology* (ed. C.J. Arthur). (London, Wishart) (written 1845–6).

MELLARS, P. 1991: Cognitive changes and the emergence of modern humans in Europe. *Cambridge Archaeological Journal* 1, 63–76.

MICHAILIDOU, A. 1999: Systems of weight and social relations of 'private' production in the Late Bronze Age Aegean. In Chaniotis, A. (ed.), *From Minoan Farmers to Roman Traders* (Stuttgart, Franz Steiner Verlag), 87–113.

MITHEN, S. 1998: The supernatural beings of prehistory and the external storage of religious ideas. In Renfrew, C. & Scarre, C. (eds.), *Cognition and Material Culture: the Archaeology of Symbolic Storage* (Cambridge, McDonald Institute), 97–106.

MORGAN, L.H. 1877: *Ancient Society*, 1958 edn. (Calcutta, Bharati Library).

POWELL, M.A. 1996: Money in Mesopotamia, *Journal of the Economic and Social History of the Orient* 39 (3), 229–42.

RENFREW, C. 1973: Monuments, mobilization and social organization in Neolithic Wessex. In Renfrew, C. (ed.), *The Explanation of Culture Change: Models in Prehistory* (London, Duckworth), 539–58.

RENFREW, C. 1974: Beyond a subsistence economy: the evolution of social organization in prehistoric Europe. In Moore, C.B. (ed.), *Reconstructing Complex Societies: an Archaeological Colloquium* (Cambridge, MA, Supplement to the Bulletin of the American Schools of Oriental Research no. 20), 69–84.

RENFREW, C. 1982: *Towards an Archaeology of Mind* (Cambridge, Cambridge University Press).

RENFREW, C. 1986: Varna and the emergence of wealth in prehistoric Europe. In Appadurai, A. (ed.), *The Secret Life of Things* (Cambridge, Cambridge University Press), 141–6.

RENFREW, C. 1996: The Sapient behaviour paradox: how to test for potential. In Mellars, P. & Gibson, K. (eds.), *Modelling the Early Human Mind* (Cambridge, McDonald Institute), 11–14.

RENFREW, C. 1997: Setting the scene. In Cunliffe, B. & Renfrew, C. (eds.), *Science and Stonehenge*, Proceedings of the British Academy 92 (London, Oxford University Press), 3–14.

RENFREW, C. 1998: Mind and matter: cognitive archaeology and external symbolic storage. In Renfrew, C. & Scarre, C. (eds.), *Cognition and Material Culture: the Archaeology of Symbolic Storage* (Cambridge, McDonald Institute), 1–6.

RENFREW, C. 2001: Symbol before concept: material engagement and the early development of society. In Hodder, I. (ed.), *Archaeological Theory Today* (Cambridge, Polity Press), 122–40.

RENFREW, C. in press: Archaeology and commodification: the role of things in societal transformation. In van Binsbergen, W. & Geschiere, P. (eds.), forthcoming.

RENFREW, C. & BAHN, P. 2000: *Archaeology: Theories, Methods, Practice,* 3rd edn. (London, Thames & Hudson).

SEARLE, J.R. 1995: *The Construction of Social Reality* (London, Allen Lane).

SHENNAN, S. 1975: The social organization at Branc. *Antiquity* 49, 279–88.

SHERRATT, A. 1997: *Economy and Society in Prehistoric Europe* (Edinburgh, Edinburgh University Press).

SHERRATT, A. & SHERRATT, S. 1991: From luxuries to commodities: the nature of Mediterranean Bronze Age trading systems. In Gale, N. (ed.), *Bronze Age Trade in the Mediterranean,* Studies in Mediterranean Archaeology 90 (Jonsered, Paul Astroms Verlag).

SPINDLER, K. 1994: *The Man in the Ice: the Preserved Body of Neolithic Man Reveals the Secrets of the Stone Age* (London, Weidenfeld and Nicolson).

THOMAS, J. 1991: *Rethinking the Neolithic* (Cambridge, Cambridge University Press).

THOMAS, J. 1996: *Time, Culture and Identity: an Interpretive Archaeology* (London, Routledge).

THOMAS, J. 1999: An economy of substances in earlier Neolithic Britain. In Robb, J.E. (ed.), *Material Symbols, Culture and Economy in Prehistory* (Carbondale, IL, Southern Illinois University), 70–89.

TILLEY, C. 1994: *A Phenomenology of Landscape* (Oxford, Berg).

VOUTSAKI, S. 1997: The creation of value and prestige in the Aegean Late Bronze Age. *Journal of European Archaeology* 5, 34–52.

WHITE, R. 1989: Towards a contextual understanding of the earliest body ornaments. In Trinkaus, E. (ed.), *The Emergence of Modern Humans* (Cambridge, Cambridge University Press), 211–31.

WHITE, R. 1993: Technological and social dimensions of 'Aurignacian age' body ornaments across Europe. In Knecht, H., Pike-Tay, A. & White R. (eds.), *Before Lascaux: the Complete Record of the Early Upper Palaeolithic* (Boca Raton, CRC Press), 277–99.

WHITTLE, A. 1996: *Europe in the Neolithic, the Creation of New Worlds* (Cambridge, Cambridge University Press).

Social Competition, Social Intelligence, and Why the Bugis Know More about Cooking than about Nutrition

JEROME H. BARKOW, NURPUDJI ASTUTI TASLILM, VENI
HADJU, ELLY ISHAK, FAISAL ATTAMIMI, SANI SILWANA,
DJUNAIDI M. DACHLAN, RAMLI, & A. YAHYA

INTRODUCTION

THE *SINE QUA NON* of a human social institution is an underlying knowledge base. Just as societies vary in their institutional structures, so too do they vary in their knowledge: one society may have great expertise in irrigation techniques, another exhaustive knowledge of local flora and fauna, another of military strategy, and yet another a highly developed theology; many societies have multiple domains of expanded knowlege. While it is apparent that social institutions such as armies and religions require elaborate knowledge bases, so too do other institutions, including kinship-based organizations (which may require genealogical knowledge or an understanding of a complex kinship system) and political-economic structures. As we shall see, even so homely an institution as the household rests on knowledge of, for example, food preparation and the nutritional needs of children.

What processes led early human societies to develop the knowledge domains that they did? Detailed histories of Neolithic and earlier societies are largely irretrievable, of course, but we can look at more recent societies and ask questions about the social processes involved in the creation of their particular knowledge bases. The two knowledge domains with which we will be concerned — cooking and child nutrition — at first glance may seem far removed from the origins of social institutions, yet they are not. The evolved psychology that both constrains and enables the generation of these two domains is very likely the same evolved psychology that permitted the development of the local knowledge that underlay even the earliest of human social institutions. Asking what social processes are involved in the selection and elaboration of knowledge bases in a contemporary society may therefore give some insight into the origins of social institutions in general.

A knowledge base may have many dimensions. However, the two with which we will be concerned are those of elaborateness/extent, and of effectiveness in the real world.

The data were collected during the course of a study of foodways and nutrition funded by the Canadian International Development Agency and coordinated by Dalhousie University. The research focused on the nutritional status of pregnant women and of children under five in two small Bugis-speaking communities in Boné District, South Sulawesi. One community is an inland, rice-growing village (Taretta), the other a coastal fishing village (Panyula). The total sample of households studied was 156. Our concern was with diet, nutrition, and food processing. The data were collected during May–August 1999 and, at the time of writing, continue to be analysed.

In the course of this study, we became aware of a certain irony: on the one hand, knowledge of cuisine was highly developed; on the other hand, knowledge of nutrition was not. An evolutionary perspective seems to suggest that the cultural emphases should have been reversed — indigenous knowledge about nutrition should have had priority over indigenous knowledge of cuisine. This chapter explores that incongruity. The plan of attack will be, first, to define some terms; second, to prepare a theoretical context for the data; third, to describe indigenous knowledge about cooking; fourth, to describe indigenous knowledge of nutrition and the nutritional status of young children and their mothers; fifth, to seek to understand how it is that people can be better cooks than nutritionists; and sixth, to take the first steps towards developing a theory of the elaboration of knowledge bases in human societies.

DEFINITIONS

We will use 'culture' to mean a pool of shared information associated with one or more populations that may be geographically localized or widely distributed or both (Barkow 1989a). It is assumed that pools overlap, and that specific items of information may occur in any number of pools. Individuals use the information in these culture pools, selecting, revising, contributing, and 'transmitting' items. A particular individual may have access to more than one pool of information: the present age of 'globalization' is one in which information pools are constantly splashing into one another. These pools of cultural information can usefully be thought of as being composed of 'particles'. By 'particles' or 'information items' within a pool is meant very loosely what some have termed 'culturgens' (Lumsden & Wilson 1981), 'memes' (Blackmore 1999; Dawkins 1976), and 'traits' (Boyd & Richerson 1985). It is not assumed that these particles are discrete — individuals constantly alter them in using them — while their 'transmission' is always problematic, involving inference and approximation rather than precise duplication (Boyer 1998; Sperber 1996). This process presumably involves various evolved mechanisms of the brain, so that different kinds of information may be processed differently (Barkow 1989a).

By 'cuisine' is meant a specific tradition of processing and preparation of food. By 'nutrition' is meant nourishment of the human body, the ingestion of nutrients that help to sustain the body in a state of health.

The term 'indigenous knowledge' is most often used in the context of socioeconomic development, where it is often associated with participatory approaches to development and with the issue of intellectual property rights. For present purposes, it will be used synonymously with the term 'local knowledge' and will refer to specific local knowledge domains. Thus, we will be speaking about indigenous knowledge of cuisine, food preparation, and child nutrition. (For useful discussions of indigenous knowledge, see Antweiler 1998; Ellen & Harris 1997; Grenier 1998; Nygren 1999; Rhoades & Bebbington 1995; Semali & Kincheloe 1999; Sillitoe 1998; Warren *et al.* 1995.)

THEORETICAL BACKGROUND

There appears to be no existing literature directly comparing the extent of indigenous knowledge of cooking with indigenous knowledge of nutrition. There also appears to be little or no literature concerning the processes whereby nutritional knowledge is generated over time, in a given society (except for studies of recent history of scientific nutrition). There is, however, a literature arguing that indigenous knowledge of food processing can increase nutritional value and/or remove toxins, and another that discusses the development of cuisines.

Adaptive food processing

We have been cooks for anywhere from 200,000 years (Brace 1996) to 1.6 million years (Wrangham *et al.* 1999: 572), presumably ample time to develop extensive knowledge of cooking. Not surprisingly, indigenous knowledge of beneficial food-processing techniques is not uncommon. Katz *et al.* (1974) found that Mayan processing of ground dried maize with limestone increases the tryptophan content while adding calcium to the diet (tryptophan being a precursor of niacin, a deficiency of which causes the disease pellagra). Bogin (1997: 117) points out that many of the foods people have eaten would be poisonous without considerable processing (e.g. manioc, horse chestnuts), while rhubarb and cashews are toxic unless treated with heat before being eaten. Spice mixtures that kill or suppress harmful bacteria and fungi are common, cross-culturally, particularly in the warm regions where they are most needed (Billing & Sherman 1998). No doubt food-processing techniques that increase availability of nutrients and/or make food safer abound (cooking, in a great many cases, being an obvious example).

Unfortunately, we do not know where this beneficial indigenous knowledge comes from. One could argue that such information particles ('memes') are invented serendipitously and then become common through processes involving 'memetics' rather than genetics or the nature of human intelligence. For example, the rule of 'imitate the successful' (Barkow 1989a; Boyd & Richerson 1985) could explain the spread of such techniques, as others noticed that the innovators and their families were healthier than most. Such conjectures may be misleading because our sample of human societies is very heavily biased — in favour of survivors! We have knowledge of successful societies only. Groups that followed practices that left them more malnourished, diseased, or poisoned than competing societies have presumably been less available for study than groups that, among other things, got at least some of their nutritional practices right. Therefore, in studying indigenous knowledge and practices regarding food, we would expect that these in general lead to proper nutrition and health, but we should not expect perfection: some practices could be harmful.

Can local food prohibitions be maladaptive?

Are cultural food prohibitions (taboos) maladaptive (or at least unhealthy) for the individuals who follow them? One school of thought argues that they can be, at least potentially (e.g. Hull 1986; Katona-Apte 1977; Wilson 1973; Wolff 1965), while others are unconvinced and/or emphasize that it is poverty that is without doubt the most important cause of malnutrition. Various studies have found that, around the world, women often have food restrictions imposed on them, either during specific periods or in general; these tend to involve high-protein foods that are mostly likely to be forbidden during pregnancy (Rosenberg 1980). For example, Gabriella Ferro-Luzzi (1973) interviewed some 1200 women in Tamil Nadu, India. She found that the women were subject to over a hundred food avoidances associated with menstruation, lactation, and childbirth, and she concluded that these restrictions were harmful to the women and to their children. Marvin Harris (1987), however, points out that the Tamil Nadu women may have regularly violated the rules, the forbidden foods may not have been part of their normal diet even when not pregnant or lactating, and they could have been compensating by eating additional foods or larger-than-usual quantities of other foods. Moreover, as a further criticism Harris points out that Ferro-Luzzi used interview data rather than systematic observation and measurement of what the women actually consumed. In support of his position, he gives us the example of Wilson's (1973) study of Ru Madu, a fishing village on the east coast of Malaysia. Wilson compared interview data on what women said they were not supposed to eat during the first 40 days post-partum with what two post-partum women actually ate during that period. There were marked discrepancies. Hull (1986) discusses ways of collecting and interpreting information about food taboos, and Laderman (1984: 547) argues convincingly for 'behavioral flexibility in the face of ideology' with regard to food behaviour.

More recent research has painstakingly focused on actual consumption and not merely claims of food restrictions. In general, the result seems to be that most of the time the restrictions make no difference, but occasionally they do. Aunger, for example, concludes that the impact of food taboos in his research area is slight, and 'only at the extreme range of undernutrition do further nutritional decrements actually translate into fitness differences' (1994a: 303). In his methodologically meticulous study (1994a, 1994b) conducted among four groups in the Ituri Forest in what is now the Democratic Republic of Congo, he found that there was evidence of lowered reproductive success (completed fertility) in only one of those groups, and then in less than 5 per cent of the women. These individuals most likely 'lived at the margin of energy balance' (1994a: 290). Kikafunda et al. (1998) found even less evidence of food taboos affecting health. They studied 261 infants/toddlers under 30 months old in Uganda. They did find much evidence of malnutrition, but anthropometric measurements showed no relationship between food taboos and either stunting or being underweight. Food taboos, in short, appear to have little or no effect on nutritional status and on genetic fitness. As Laderman (1984: 549) points out, the term itself seems to imply a rigid rule and a belief in supernatural repercussions if it is transgressed. In fact, as she discovered for a Malay village in Malaysia and as we found in South Sulawesi, food 'taboos' are better thought of as rules that may be interpreted flexibly or simply ignored. Moreover, for few if any societies do such prohibitions make up more than a small part of indigenous knowledge and practices regarding food and nutrition.

Two caveats are necessary before we begin describing indigenous knowledge of cuisine and nutrition in the two South Sulawesi communities studied. First, as we have already argued, it is a serious error to assume that people rigidly follow their food ideologies. Second, as we shall shortly see, the communities studied have not so much a system of 'food prohibitions' as a complex and varyingly known balance theory relating food, bodily state, and health to one another. Such a conceptualization should not be reduced to taboos or prohibitions, as the system as a whole may encourage the consumption of some foods even as it discourages the consumption of others, and may be interpreted differently from family to family.

CUISINE (INDIGENOUS KNOWLEDGE OF COOKING) IN THE STUDY VILLAGES

The simplest means of establishing the elaborateness and sumptuousness of Bugis cuisine would be to serve a meal typical of celebrations (Bug. *pesta*). Perhaps in the future, multimedia presentations will permit virtual meals, but for the moment we must dine on description. Paul Rozin's (1987) concept of 'flavour principle' is useful here. Rozin pointed out that national cuisines tend to have distinctive flavours due to the use of certain ingredients. For Chinese food, Rozin informs us, the ingredients are ginger, rice wine, and soy sauce; for Mexican food they are chilli peppers and lime and/or tomato. For Bugis food, the research team — which included several individuals expert in Bugis cuisine, and who consulted friends and family members — determined that the distinctive flavour principle is derived from the following ingredients, in order of priority:

- candlenut
- onion
- tamarind
- white pepper

Additional ingredients frequently found in Bugis cuisine are coconut milk, chilli peppers, and *terasi* (fermented fish or shrimp paste; *terasi* is Indonesian, *tarasi* is Bug.). Bugis food in part symbolizes Bugis social identity, and *lawa* (Bug. chopped raw fish or vegetable and grated coconut, seasoned and mixed with an acid such as lime juice) is the dish that my Bugis colleagues agree is especially 'Bugis'. (Not surprisingly, informants from the fishing village studied particularly prized *lawa bale*, fish lawa.)

Bugis people are not the only ones who enjoy Bugis food. Christian Pelras (1996: 228), author of a comprehensive study of the Bugis kingdoms, describes how one James Brook, visiting the region in 1840, was singularly impressed, while Pelras himself writes about 'the excellence and delicacy of Bugis cooking, which can be experienced not only among aristocratic families but also in very simple and even poor households' (1996: 22). He adds that 'sweets and pastries . . . are produced in innumerable variety' (228). Susan Millar (1989), in her tightly focused study of weddings and status in another Bugis district, Soppéng, discusses the immense amount of care and labour involved in the preparation of the food served at a Bugis wedding and the importance for determining the hosts' status of the foods' variety, quantity, and quality. (Today, in the more urban areas, families are likely to hold the wedding at a restaurant, substituting money for the clients who, in more rural areas and in the past, would have done the wedding cooking; even in rural areas, some families will now hire a professional caterer.)

Below are a number of recipes, collected in the study villages by Dr Elly Ishak, which may give some sense of the nature of Bugis cooking.

Bugis recipes

Masak santan (mixed vegetables in coconut milk)
Vegetables (immature jackfruit, pumpkin, eggplant) are cooked in coconut milk with turmeric, onion, chilli peppers, and lemon grass.

Beppa janda
A wrapper is made from finely grated cassava and salt. It is rolled into a tube around a banana. The dish is steamed, then served sprinkled with grated coconut. (*Beppa janda* is eaten both as a snack or as part of a lunch or supper.)

Nasu likku (coconut cream chicken)
Chicken is cooked with coconut milk and onion, garlic, candlenut, caraway seed, pepper, galangal, laurel leaf, lemon grass, palm sugar, and salt. *Nasu likku* is served either with steamed rice or with boiled rice cake (Bug. *burasa*).

Bale tapa (fish smoked over a grill)
Ground candlenut, chilli, garlic, onion, and soy sauce are mixed together with fresh fish (usually milkfish), which is then either roasted or grilled.

Nasu bale (stewed fish).
The most commonly used fish for this dish is fresh *cakalang*, a type of tuna. The fish is thoroughly cleaned, then cut into several pieces and washed until all the blood has been removed. It is put in a pot together with water, turmeric, onions, tamarind juice, monosodium glutamate, and salt. The pot is brought to a boil, then simmered for 45–60 minutes.

No claim is here made that these and other Bugis dishes are unique, or that other Indonesians would consider Bugis cooking superior to that of their own home regions — Indonesia is a land of notable regional cuisines, after all, and there are certainly marked similarities between Bugis food and, for example, the Malay dishes described by Wilson (1986). The only point of this discussion is to establish that the women interviewed (cooking is gendered, with men being in principle forbidden even to enter a kitchen, though they may grill some foods in the garden or aboard their fishing boats) in the two study areas had extensive knowledge of and skill in cooking, permitting us the generalization that the knowledge domain of cuisine is well developed and highly effective in the local culture.

UNHEALTHY LOCAL KNOWLEDGE OF NUTRITION: SOME LIKELY SUSPECTS

It is important to make clear at the outset what is *not* being argued. First, it is not being argued that the major cause of malnutrition in the two communities is indigenous knowledge of nutrition or food ideology or lack of education about scientific nutrition; poverty is no doubt the major cause of malnutrition in both populations. Second, it is not being argued that diet is the sole cause of malnutrition; parasitic infections (e.g. nematodes) and other diseases may also play a role, but, as no data on their prevalence were collected, they will not be discussed. Third, no implication is intended that the food beliefs and lack of nutritional knowledge prevalent in the study communities are unique; as we will see, it is the *lack* of uniqueness that gives the data their theoretical import. Fourth, while the focus of the discussion is on food beliefs and practices likely to be contributing to malnutrition in the study communities, it is not being argued that such practices are typical; a great many local food practices and beliefs no doubt make for healthy eating. For example, drinking water was invariably first boiled and then kept in covered containers in the study villages, as has long been the practice in the Bugis communities of South Sulawesi (Pelras 1996). What *is* being argued is simply that local knowledge and practice in some cases appear to contribute to malnutrition.

The clearest example the research team found of a local food belief and practice likely to have a negative impact on nutrition was denying the newborn the mother's colostrum. (Colostrum is the pre-milk breast secretion; the actual milk does not appear until the second or third day after birth, or even later.) Colostrum is quite important for the health and nutrition of the infant (Barkow & Hallett 1989). Not only does it provide the neonate with sterile fluid, it also permits it in effect to share the mother's immune system until its own has matured somewhat. Popular books today focus on the health benefits of colostrum (e.g. Hawken 1999; Ley 1997), while current animal research finds that it not only provides immunological advantages but also increases the general vitality of the young animal (Blum *et al.* 1997; Burrin *et al.* 1997; Hadorn *et al.* 1997). Key informants in both study communities (including ritual practitioners/healers (Ind. *dukun*, Bug. *sanro*) and older women in general) agreed that, in the past, infants would be given the breast only on the third day so as to avoid the colostrum. One woman, telling us that with her later children she had followed the health post's advice to give the breast immediately, remarked that doing so had been distasteful and difficult. Colostrum denial was found in approximately one-third of the Human Relations Area Files societies for which data were available (Barkow & Hallett 1989: 305). The practice has in the past been common in Indonesia and in Thailand (Van Esterik 1989: 129), but today it is waning due to the influence of education, the

local health posts and (in the case of Indonesia) government training programmes for *dukun*; it remains common in the study villages, though precise
data were not collected. It seems possible that colostrum denial is to some
degree responsible for the high rate of morbidity in the communities studied: of
the 199 children aged 0–60 months in the two study samples, approximately 25
per cent were described by caregivers as having been ill during the three weeks
prior to the interview.

A second idea that appears likely to have nutritional impact in the two study
communities is the belief that pregnant women should eat little, and especially
not 'hot' foods (which tend to be the high-protein foods). (As was discussed
earlier, similar prohibitions have been very common in much of the world.)
This was regularly explained in terms of avoiding having a large infant who
might be difficult to deliver and whose birth might tear the perineum. Two
health-centre midwives interviewed in the farming village of Taretta explained
that women in the early stages of pregnancy often believed that they should not
drink milk, and that in the later stages they should avoid meat, beans, and
peanuts. (They do, however, believe that it is important that they eat vegetables.)
Lack of proper maternal nutrition during pregnancy can result in a low birthweight infant (that is, under 2.5 kg at birth); low birthweight babies are susceptible to neurodevelopmental disorders, including cerebral palsy, and may suffer
from poor health later in life. For Taretta, some 33.3 per cent of the 97 infants
in the study sample had low birthweight; for Panyula, the comparable figure
was 11.3 per cent of 102 infants. Because low birthweight is not necessarily
caused by poor maternal nutrition (pre-term delivery is a possible cause, for
example), the data can only be considered suggestive; the belief that pregnant
women should eat little, especially of high-protein foods, may be having a negative effect on foetal growth and infant health.

A third belief (similar to that described by Laderman (1984: 553) for a
Malaysian village) is that children will develop a parasitic infection — usually
described as 'worms' (Bug. *cacingan*) — if fed too much fish; the stomach is
said to swell and 'makes noises'. (One informant, in explaining this belief,
quickly added that 'now we know that it is caused because their hands are not
clean'.) Survey data collected and an analysis of the nutritional adequacy of
the diet (based on a 24-hour dietary recall survey) shed some light on young
children's fish consumption. Dietary recall data and anthropometric measurements are only snapshots which refer to a single point in time and do not reveal
possible seasonal variation, and the present study could only collect data at a
single point. However, Table 1 shows that of the 23 Taretta children aged 12–23
months old in the sample, 30.4 per cent were never given fish to eat. For
Panyula, of the 27 children of that age in the sample, 18.5 per cent did not
receive fish. If we move to the 24–60 month age group (Table 2), these figures
change: for Taretta, 57.7 per cent of the 52 children in the sample received fish

Table 1. Percentage of children 12–23 months given specific foods in Taretta (n = 23) and Panyula (n = 27).

Type of food	Never		1–3 times per month		1–2 times per week		3–6 times per week		Every day	
	Taretta	Panyula	Taretta	Panyula	Taretta	Panyula	Taretta	Panyula	Taretta	Panyula
Infant formula	95.7	92.6	0.0	0.0	4.3	0.0	0.0	0.0	0.0	7.4
Processed food	100.0	100.0	0.0	0.0	0.0	0.0	0.0	0.0	0.0	0.0
Other milk	82.6	77.8	0.0	14.8	4.3	3.7	8.7	0.0	4.3	3.7
Meat (beef/goat/etc.)	56.5	77.8	43.5	18.5	0.0	3.7	0.0	0.0	0.0	0.0
Fish	30.4	18.5	4.3	0.0	0.0	7.4	17.4	22.2	47.8	51.9
Poultry	43.5	77.8	52.2	22.2	4.3	0.0	0.0	0.0	0.0	0.0
Liver (beef/chicken)	61.9	79.2	38.1	20.8	0.0	0.0	0.0	0.0	0.0	0.0
Eggs	28.6	16.7	14.3	16.7	23.8	37.5	33.3	29.2	0.0	0.0
Tempeh/tofu	76.2	66.7	9.5	16.7	9.5	12.5	4.8	4.2	0.0	0.0
Mung bean/peanut	9.5	25.0	4.8	25.0	19.0	12.5	33.3	29.2	33.3	8.3
Green leafy vegetables	4.8	16.7	4.8	12.5	4.8	16.7	19.0	25.0	66.7	29.2
Red/yellow vegetables	38.1	50.0	42.9	16.7	14.3	20.8	4.8	8.3	0.0	4.2
Red/yellow fruits	33.3	4.2	14.3	33.3	38.1	20.8	4.8	33.3	9.5	8.3
Other fruits	9.5	4.2	4.8	0.0	42.9	25.0	28.6	50.0	14.3	20.8
Fruit juice	100.0	100.0	0.0	0.0	0.0	0.0	0.0	0.0	0.0	0.0
Snack	14.3	16.7	9.5	0.0	38.1	20.8	23.8	29.2	14.3	33.3

Table 2. Percentage of children aged over 24 months given specific foods in Taretta (n = 52) and Panyula (n = 47).

Type of food	Never		1–3 times per month		1–2 times per week		3-6 times per week		Every day	
	Taretta	Panyula	Taretta	Panyula	Taretta	Panyula	Taretta	Panyula	Taretta	Panyula
Infant formula	98.1	97.9	1.9	0.0	0.0	0.0	0.0	0.0	0.0	0.0
Processed food	100.0	100.0	0.0	0.0	0.0	0.0	0.0	0.0	0.0	0.0
Other milk	55.8	31.9	7.7	25.5	26.9	21.3	3.8	4.3	5.8	17.0
Meat (beef/goat/etc.)	65.4	61.7	28.8	34.0	5.8	2.1	0.0	0.0	0.0	0.0
Fish	0.0	0.0	5.8	0.0	9.6	4.3	26.9	6.4	57.7	89.4
Poultry	48.1	42.6	46.2	53.2	3.8	4.3	1.9	0.0	0.0	0.0
Liver (beef/chicken)	69.2	54.2	30.8	41.7	0.0	4.2	0.0	0.0	0.0	0.0
Eggs	12.8	0.0	23.1	41.7	33.3	33.3	28.2	12.5	2.6	12.5
Tempeh/tofu	82.1	45.8	10.3	37.5	2.6	12.5	5.1	0.0	0.0	4.2
Mung bean/peanut	5.1	20.8	7.7	33.3	25.6	12.5	30.8	25.0	30.8	8.3
Green leafy vegetables	2.6	4.2	2.6	4.2	17.9	25.0	30.8	37.5	46.2	29.2
Red/yellow vegetables	48.7	37.5	35.9	33.3	12.8	20.8	2.6	4.2	0.0	4.2
Red/yellow fruits	17.9	16.7	12.8	33.3	41.0	25.0	25.6	20.8	2.6	4.2
Other fruits	0.0	0.0	2.6	0.0	25.6	62.5	66.7	29.2	5.1	8.3
Fruit juice	100.0	100.0	0.0	0.0	0.0	0.0	0.0	0.0	0.0	0.0
Snack	2.6	0.0	2.6	0.0	17.9	25.0	59.0	33.3	17.9	41.7

daily, while for Panyula the comparable figure was some 89.4 per cent of the 47 sample children. In short, children often are not given fish to eat on a regular basis until they are two years of age. (Not surprisingly, they are given fish more often in the fishing village of Panyula than in the farming village of Taretta.) Does the presumed reduction caused by the 'fish causes worms' belief in the frequency and amount of fish given the children result in protein deficiency?

Tables 3a and 3b show that the protein intake of infants under two is not inadequate. The two- to five-years-old group for Panyula receives 98.3 per cent of the recommended daily amount (RDA), while that age group in Taretta receives 73 per cent RDA. Thus, beliefs notwithstanding, Panyula children are receiving adequate protein, and Taretta children only somewhat less. While one could argue that the 'fish causes worms' belief does lower the protein intake of the children of Taretta, more likely factors involve the fact that, first, fish are far less available in the inland community of Taretta than in the coastal fishing village of Panyula; second, the market of Taretta meets only every fifth day while that of Panyula meets every day, affecting food availability (particularly that of

Table 3a.   Quality of nutrient intake in Taretta (n = 86).

Variable	6–11 months[a]			12–23 months[a]			≥ 24 months[b]		
	X[c]	RDA	%RDA	X	RDA	%RDA	X	RDA	%RDA
Energy	132.0	269.0	48.9	259.0	746.0	34.7	449.0	1250.0	35.9
Protein	4.5	2.0	225.0	8.5	5.0	170.0	16.8	23.0	73.0
Vitamin A (RE)	44.3	13.0	340.8	99.6	126.0	79.0	79.3	350.0	22.7
Vitamin D (RE)	0.6	6.6	9.1	2.1	7.0	30.0	3.9	10.0	39.0
Vitamin B1 (mg)	0.0	0.1	0.0	0.1	0.4	25.0	0.2	0.5	40.0
Vitamin B2	0.1	0.2	50.0	0.1	0.4	25.0	0.1	0.6	16.7
Niacin	0.3	3.0	10.0	1.0	7.0	14.3	2.4	5.4	44.4
Vitamen B6	0.1	0.0	0.0	0.1	0.0	0.0	0.3	0.0	0.0
Pantotenic acid	0.5	0.5	100.0	0.6	0.7	85.7	1.1	0.0	0.0
Folate (mcg)	14.0	0.0	0.0	21.0	3.0	700.0	29.9	40.0	74.8
Vitamin B12 (mcg)	0.3	0.0	0.0	0.7	0.0	0.0	1.1	0.5	220.0
Vitamin C (mg)	1.4	0.0	0.0	1.9	8.0	23.8	5.1	40.0	12.8
Zinc (mg)	0.5	4.2	11.9	0.7	5.8	12.1	1.3	10.0	13.0
Iron (mg)	0.4	10.8	3.7	0.8	5.8	13.8	1.2	8.0	15.0
Magnesium (mg)	14.2	51.0	27.8	27.5	66.0	41.7	54.0	0.0	0.0
Sodium (mg)	30.1	199.0	15.1	48.0	401.0	12.0	72.6	0.0	0.0
Phosphorous (mg)	71.0	306.0	23.2	125.9	193.0	65.3	220.9	250.0	88.4
Calcium (mg)	28.2	336.0	8.4	43.5	196.0	22.2	39.4	500.0	7.9
Potassium (mg)	107.7	346.0	31.1	196.5	512.0	38.4	367.2	0.0	0.0
Copper (mg)	0.0	0.1	0.0	0.1	0.3	33.3	0.2	0.0	0.0

[a]RDA taken from WHO (1998) after adjusting for breastmilk intake.
[b]RDA taken from WHO (1998).
[c]X refers to 'mean amount'.

a food as perishable as fish); and, third, Taretta as a whole is less prosperous than is Panyula. It thus seems unlikely that the belief that fish cause worms in young children has any real nutritional impact in either village (a conclusion similar to that of Laderman (1984) for the same belief's effect in the Malaysian community that she studied). Even if the belief does affect the amount of fish given to the children, in both communities the two- to five-year-old group has other protein sources, including tofu and tempeh, mung beans and peanuts, and eggs.

Balance theory

Both study communities shared a version of balance or hot/cold theory that is similar though not identical to indigenous systems described for other areas of south-east Asia (e.g. those described for peninsular Malaysia by Manderson (1986), for Malays by Wilson (1973) and for Javanese by Hull (1986)). Many individuals interviewed described foods and illnesses as being either hot or cold, and the body itself as often being in either a hot or cold state. (However,

Table 3b. Quality of nutrient intake in Panyula (n = 87).

Variable	6–11 months[a]			12–23 months[a]			≥ 24 months[b]		
	X[c]	RDA	%RDA	X	RDA	%RDA	X	RDA	%RDA
Energy	173.0	269.0	64.5	340.0	746.0	45.5	520.0	1250.0	41.6
Protein	7.1	2.0	355.0	12.5	5.0	250.0	22.6	23.0	98.2
Vitamin A (RE)	44.2	13.0	340.0	121.4	126.0	96.3	130.9	350.0	37.4
Vitamin D (RE)	2.6	6.6	39.4	5.0	7.0	71.4	9.7	10.0	97.0
Vitamin B1 (mg)	0.1	0.1	100.0	0.1	0.4	25.0	0.2	0.5	40.0
Vitamin B2	0.1	0.2	50.0	0.1	0.4	25.0	0.2	0.6	33.3
Niacin	0.8	3.0	26.7	1.8	7.0	25.7	3.1	5.4	57.4
Vitamen B6	0.1	0.0	0.0	0.0	0.0	0.0	0.0	0.0	0.0
Pantotenic acid	0.6	0.5	120.0	0.9	0.7	128.6	1.6	0.0	0.0
Folate (mcg)	12.3	0.0	0.0	23.9	3.0	796.7	35.3	40.0	88.3
Vitamin B12 (mcg)	0.7	0.0	0.0	1.2	0.0	0.0	2.3	0.5	460.0
Vitamin C (mg)	1.8	0.0	0.0	5.3	8.0	66.3	6.4	40.0	16.0
Zinc (mg)	0.5	4.2	11.9	0.9	5.8	15.5	1.6	10.0	16.0
Iron (mg)	0.5	10.8	4.6	0.9	5.8	15.5	1.4	8.0	17.5
Magnesium (mg)	18.0	51.0	35.3	39.6	66.0	60.0	62.0	0.0	0.0
Sodium (mg)	30.1	199.0	15.1	47.0	401.0	11.7	70.9	0.0	0.0
Phosphorous (mg)	114.8	306.0	37.5	188.5	193.0	97.7	342.6	250.0	137.0
Calcium (mg)	28.8	336.0	8.6	33.1	196.0	16.9	46.8	500.0	9.36
Potassium (mg)	147.2	346.0	42.5	336.7	512.0	65.8	518.7	0.0	0.0
Copper (mg)	0.0	0.1	0.0	0.1	0.3	33.3	0.2	0.0	0.0

Notes: see Table 3a.

some illnesses, explained one *sanro*, were due to supernatural causes.) But knowledge of this theory was not evenly distributed. During one group interview, as a 70-year-old *sanro* explained the system, two married women in their twenties listened with expressions of surprise on their faces. They afterwards explained that they had had little knowledge of the system as a whole, though they were aware that there were foods that pregnant women should not eat, foods for particular illnesses, and so forth. Older women generally had a greater awareness of the system than did younger women. (No men were interviewed on this topic.)

Focus groups and key informants from both communities agreed on what might be thought of as 'core' indigenous knowledge of nutrition. All those interviewed agreed that what was tasty and filling was good for the body, and that rice was by far the most important food — even the fishers agreed that rice was more important than fish, the second food that would be mentioned. ('You can always eat rice with salt, but with fish, you don't have a meal,' explained one man.) They also tended to agree that some foods were 'hot' and some were 'cold', as were some body states (e.g. pregnancy and menstruation are hot). Tables 4 and 5 summarize the consensus. There are degrees of being 'hot' and 'cold' both for people and for foods. A pregnant women is hot (ice is believed to cause miscarriage). A woman in labour is very hot and remains so after delivery, until (according to one *sanro*) her true milk comes in and she becomes cold (and

Table 4. Foods frequently described as hot and as cold.

Hot foods	Cold foods
beef	papaya
pineapple	sweetsop (Ind. *sarakaya*)
jackfruit	bananas
tiger mango (Bug. *pao maccan*)	chicken
goat	boiled rice
sticky rice	peanuts
sambal (*sambal* is Ind.; *peco ladang* is Bug;	mung beans
the term refers to a sauce made of chilli	string beans
pepper, tomato, and *terasi*)	eggplant
horsemeat	kelor (the pinnate leaves of the merunggai tree,
palm sugar	eaten as a vegetable and considered to be
mango	especially cold — people are said to shiver
fermented cassava or rice (*tape*)	after eating kelor)
fish	cassava
salt fish	yam
dried fish	breast milk
durian	cucumber
ginger	watermelon
	greater galangale (galangal)
	immature coconut

Table 5.   Body states.

Hot	Cold
men	women
parturient women	infants
lactating women	
menstruating women	

therefore should eat hot foods to restore balance). A lactating woman is hot but milk is cold. Men are hotter than women. Meat is hot, as are the larger fish such as tuna. Ocean fish are hotter than freshwater fish. Two older key informants felt that eating hot food makes people more emotional, more easily angered and hot-tempered. People also tend to feel physically hot when they eat hot foods, and cold when they eat cold food. Older women explained that the hot foods they had avoided when younger now gave them no trouble (such as headache). When asked about pregnancy cravings, women and men occasionally listed hot foods that wives had craved and that husbands had done their best to supply, regardless of the hot state of pregnancy. There was agreement that if one was in a cold state then an especially hot food should be avoided, though (according to some informants) a food that was only somewhat hot would be safe, and in some instances a cold food would actually be advisable. One *sanro* explained that papaya, being cold, was good for fever. A lactating woman, being hot, should avoid palm sugar and other hot foods, while ensuring a good (cold) milk supply by eating cold foods. Healers varied in the details of their accounts.

Many specific food beliefs held by some individuals appear to have little to do with hot/cold theory but much to do with the kind of similarity of cause-and-effect found in 'sympathetic magic' and in homeopathic medicine. Thus, cucumber and papaya are not good for girls because they are watery and this will make the girl's vagina too wet and displease her eventual husband. Men should eat bamboo shoots and the head of the 'gold fish' (*Cyprinus carpio*) because these improve virility. (Eggplant, however, is said to cause male impotence.) Eating the 'giant squid' (Bug. *gurita*) will cause a pregnant woman to have a difficult delivery. Similarly, a pregnant woman should not eat seaweed lest she suffer the medical condition *mola hidatidosa* (Bug. *hamil anggur*, literally, 'pregnant grapes', in which the apparent pregnancy is due to the uterus being filled with grape-like growths). Coconut milk and oil make for an easy delivery, as does having the woman in labour take a mouthful of water and then spit it out. Pineapple can cause a miscarriage. A nursing mother must not eat banana blossom (which, as the fruit itself grows, appears to shrink, and is thought to cause the infant to shrink as well). A father-to-be must not kill any animal, though fishing is permitted. He must not eat duck lest his child be born with webbed digits, and if he opens a water gate to flood his paddy his son will be born with a cleft palate. (The *sanro* interviewed also considered many foods to be curative for specific diseases, but these will not be listed here, in the interests of brevity.)

Food storage and processing

While anthropologists have traditionally collected 'beliefs', they have rarely looked at the consequences for nutrition of food storage, processing, and cooking techniques. Two members of the health team, Elly Ishak (a food technologist) and Faisal Attamimi (a pharmacologist and toxicologist), undertook work to address this gap. In general, they found no variation in basic cooking and storage techniques in the two communities, and little variation from household to household. Several food-processing practices, in their opinion, lowered the nutritional value of foods. Rice, a staple food, would be washed from three to six times in copious amounts of water. Unfortunately, this practice tends to remove the water-soluble vitamins; from a nutritional perspective, a single washing would be preferable. Vegetables would be washed, sliced, and then left to soak for long periods, presumably reducing much of the vitamin C and thiamine content. Vegetables would then be cooked for about 30–40 minutes, until very soft, resulting in additional vitamin loss. Earlier, the ingredients for making stewed tunafish were listed. The fish is simmered for 45–60 minutes. From a nutritional point of view a very brief cooking time would be preferable, as long cooking damages the protein and lowers its nutritional value.

Malnourishment, beliefs, and food processing

Table 6 shows that, among children under five years old, malnourishment in the two study villages ranges from about 35 per cent to 52.5 per cent (depending on village, type of measurement, and gender). (See Tables 3a and b for the nutritional analysis of the dietary recall survey.) As for the mothers of these children, Table 7 shows that, as assessed by standard anthropometric measures, 41.5 per cent of them are malnourished.

It is impossible to determine, from these data, the precise contribution to malnutrition (if any) made by beliefs about foods and by the techniques used in preparing food. Suppose, however, that poverty is the sole factor in the malnourishment the study documented; if so, then we should expect to find fewer malnourished infants and young children among those who are more prosperous than among those who are less prosperous.

The relationship between nutritional status indicators and socioeconomic indicators is shown in Table 8. The data were collected by Drs Nurpudji and Veni Hadju (both physicians and faculty members at Hasanuddin University

Table 6.   Prevalence of malnutrition in children under five years by sex.

Nutritional status	Taretta (%) n = 102	Panyula (%) n = 97	Total (%) n = 199
Underweight			
boys	38.1	35.1	36.4
girls	38.3	42.5	40.0
Stunting			
boys	31.0	36.8	34.3
girls	35.0	52.5	42.0
Wasting			
boys	9.5	10.5	10.1
girls	5.0	5.0	5.0

Table 7.   Nutritional status of mother.

Nutritional status	Taretta (%) n = 87	Panyula (%) n = 72	Total (%) n = 159
Arm circumference			
malnourished	48.3	33.3	41.5
normal	51.7	66.7	58.5
Body mass index (BMI)			
< 18.5	16.1	9.7	13.2
18.5–25	67.8	69.4	68.6
> 25	16.1	20.8	18.2

(UNHAS)). They used several measures of malnutrition: HAZ refers to height for age, WAZ weight for age, and WHZ weight for height (the Z stands for 'z-score'). The norms used are those considered standard by the World Health Organization. The indicators of socioeconomic status they used are self-explanatory, with the exception of the 'poor' category, which had to do with proportion of income spent for food. Families spending 70 per cent or more of their income on food were categorized as 'poor'; those spending a lower proportion of income were labelled 'not poor'. Only the relationship between WAZ (weight for age) and father's occupation achieves statistical significance (p=.04), though the relationship between WAZ and the 'poor' category is marginally significant (p=.076). HAZ (height for age) and WHZ (weight/height ratio) measures are not significantly associated with any of the socioeconomic indicators. Similarly, size of family, mother's education, and presence/absence of a television were not associated with any measures of malnutrition. It would appear that food beliefs/practices, and not just poverty, may contribute at least somewhat to infant and child malnutrition in the two communities studied. (An alternative interpretation is that those with slightly greater income choose not to use it to provide additional food for their young children. It would be interesting to replicate this study in communities with a larger proportion of genuinely prosperous families, using a larger total sample.)

Table 8.   X-square significance between nutritional status and socioeconomic variables.

	WAZ-score		HAZ-score		WHZ-score	
	Normal	Underweight	Normal	Stunted	Normal	Wasting
Size of family						
< 5	42	26	40	28	64	4
6–10	50	28	49	29	72	6
> 10	4	3	5	2	6	1
p	ns		ns		ns	
Poor category						
Yes	76	51	76	51	118	24
No	20	6	18	8	9	2
p	0.076		ns		ns	
Mother's education						
≤ primary school	59	37	57	39	90	6
above primary school	37	19	37	19	90	6
p	ns		ns		ns	
Father's occupation						
farmer/fisher, labourer	56	42	57	41	91	7
Government official, own business	38	14	35	17	48	4
p	0.040		ns		ns	
Television						
Yes	34	18	33	19	50	2
No	62	38	60	40	92	8
p	ns		ns		ns	

DISCUSSION

The two domains of Bugis knowledge studied — cuisine and infant/child nutrition — both seem to be elaborate and extensive. They differ, however, in effectiveness. Bugis cuisine, if consensus about its quality is considered worthy data, is highly effective in producing many well-liked dishes. Bugis child nutrition, however, is rather haphazard, with practices that may be harmful to infants (particularly infants made vulnerable by other circumstances, such as illness). In the case of colostrum denial, it is very likely that this practice is harmful. Moreover, the preparation of vegetables and the long cooking of fish may also be contributing to malnutrition. That the less poor generally have children as malnourished as the more poor supports the real possibility that food beliefs and practices may be contributing to malnutrition in the study villages.

The thought here is not that Bugis food practices are particularly harmful or unique. It is important to keep in mind that many of the food practices and beliefs the research team found are more likely to be healthful than maladaptive; for example, in a world with little access to Caesarean sections (as was the case, historically), efforts to avoid large babies, in spite of the risk of low birthweight infants (and all the sequelae attached), may actually increase completed fertility rates. Then, too, it may be that the danger of very young children choking on the small bones of fish may outweigh the benefit of additional protein. Many of the food proscriptions and prescriptions for particular bodily states could have unknown but nevertheless real health benefits. Moreover, the lack of systematic knowledge of nutrition is anything but unusual — historically, it has been true of all of the world's peoples.

The point of this discussion is not to dwell on the shortcomings of Bugis child-feeding practices, but to seek explanation for how it is that the same people who have a superbly effective knowledge of cooking have a very imperfect child-feeding knowledge domain. Why should this be so? Why is it that nutritionally wise beliefs and practices are not universal? In the two study communities, how is it that the indigenous knowledge of cooking that the women share permits them to prepare impressive and delicious feasts, while their indigenous knowledge of child nutrition is, at best, spotty, at worst, somewhat harmful? Why have they developed more expertise in cooking techniques than in child feeding? Why, in their food processing and cooking, have they not invented a set of nutrient-sparing techniques? It is as if, over historical time, far more thought and energy have been spent in perfecting a lovely cuisine than in determining how best to feed children! During the hundreds of thousands of years that we have been using fire, we have often become expert cooks, but we have usually remained relatively inept nutritionists.

Let us try to produce a provisional answer to this conundrum, so that we can return to our initial focus on the origins of the knowledge foundations of social institutions. A good place to begin is with the nature of human intelligence and its relationship to socially transmitted information.

Human intelligence in evolutionary perspective

Human intelligence seems to have evolved in part to keep our excessive reliance on socially transmitted information from having maladaptive consequences (Barkow 1989a, 1989b), and in part to solve problems of social living, including social competition (Byrne & Whiten 1988; Humphrey 1976, 1983; Whiten & Byrne 1997). (For an alternative view — that human 'creative intelligence' is largely the product of sexual selection — see Miller (2000).)

Our evolutionary history has left us with a hypertrophied reliance on socially transmitted information. Especially as children, we are deeply dependent on the pools of transmitted information we loosely call 'culture', pools that include, of course, indigenous knowledge. But this is a risky adaptive strategy: socially transmitted information can at times be ineffective or even maladaptive. Various processes (discussed at length in Barkow 1989a, 1989b) lead to this situation, as when ecologies alter so that formerly adaptive practices turn out to

have negative long-term consequences (e.g. climate change, overfishing, population growth, etc.), or when uncorrected errors gradually accumulate over
time (e.g. 'colostrum is bad for the infant'). However, the most important
source of 'bad' socially transmitted information, for present purposes, is a
byproduct of social competition: self-interested bias. We tend to invent, edit,
and revise socially transmitted information in ways that support our own interests both as individuals and as groups (e.g. 'our religion/ethnic group/class/
academic discipline, etc. is superior to all the others').

This tendency for socially transmitted information to contain potentially
maladaptive information has apparently led to selection for various error-
correction mechanisms. For example, the problem of accumulating error seems
to be dealt with in part by so-called 'adolescent rebellion', which (whatever its
other functions) appears to serve partly as a general editing mechanism in
which much that parents and others have sought to transmit to the young
person is called into question, challenged, and compared with alternative
particles of information (Barkow 1989a, 1989b). Another way of dealing with
the problem of accumulating error may be our tendency to attend preferen-
tially to high-status individuals and learn from them, rather than from those
lower in status, thereby eliminating some less-than-adaptive practices in favour
of some that have a higher probability of effectiveness in the real world
(Barkow 1989a). Given that social status tends to be tied to genetic fitness (e.g.
Cronk 1991), these high-ranking information sources are likely to be a better
source of effective particles of culture than the low-ranking.

Perhaps the most powerful mechanisms for correcting socially transmitted
information are those that track the self-interest of others in their roles as
suppliers of information. If you are suspicious of the arguments we are making
in this chapter, then you are exemplifying one of these mechanisms: distrust!
We are often suspicious of the information conveyed by individuals and groups
with 'vested interests', and we seek to evaluate their motives and biases and
past record of accuracy. Suspicion also plays a role in the regulation of social
exchange.

Social exchange is a core component of human society, a component that
in the context of biological evolution is often referred to as 'reciprocal altruism'.
Here we have another risky adaptive strategy — the benefits of mutual
exchange of resources and of aid in general are obvious, but so is the risk of
being cheated. One protective mechanism that appears to have evolved as a
result is the specialized memory involved in reciprocal altruism. We readily
recall every instance in which we have aided another, especially when the aid
may not have been fully reciprocated (Cosmides & Tooby 1992); we tend selec-
tively to forget instances in which others have aided us, especially those in which
we ourselves never fully repaid the altruist: many a divorce case shows how
effective our biased memories can be.

Another example of an aspect of human cognition that reflects the social nature of our evolutionary environment is that of gossip. Consider, for example, the strength of our interest in the sexual and status-related activities of high-ranking members of our local group. This interest is what underlies the trans-cultural universal gossip (Barkow 1992) and the modern phenomenon of the soap opera.

Consistent with the idea of social intelligence, we note that social competition serves to focus and even enhance our intelligence and problem-solving ability; when the competition is group competition, it increases our ability to co-operate. Knowledge domains related to competition will probably be the most elaborate and effective sectors of the cultural pool.

Finally, social competition and informational editing involve a rather brief timeframe during which feedback from physical and social events is evaluated. We seem adept at solving problems in which there is feedback in minutes and we are often successful even when feedback requires months. We are not very good at problems involving much longer timeframes, however.

Now we have some background theory about the nature of human intelligence and cognition and the various evolved mechanisms that serve to limit the amount of socially transmitted error in our cultural information pools. We are now almost ready to return to the core question of why Bugis knowledge of cuisine seems to be more effective than their knowledge of nutrition.

Human intelligence and indigenous knowledge

The discussion of the evolved psychology of human intelligence suggests that, for any domain of indigenous knowledge, one can ask the following questions:

- Is social competition involved in knowledge production?
- Does self-interest play a role in disseminating and revising information?
- Are the problems comparable to the difficulties of social living our intelligence arguably evolved to solve, particularly with respect to time-frame?
- What error-correction mechanisms are relevant? Are there specialized mechanisms primarily applicable to this domain?

Now, let us ask these questions of the Bugis domains of cuisine and of child feeding/nutrition practices.

Bugis cuisine

Bugis cooking historically involved competition. As was earlier discussed, at the Bugis *pesta* (celebrations), the hosts' status depends in part on the quantity and quality of the food produced. Susan Millar (1989) describes, for the neighbouring District of Soppéng, the importance of weddings and other *pesta* in determining relative social position, and how quality and quantity of food (cooked by clients) plays an important role in establishing or confirming the relative standing of families. It is likely that, for centuries, Bugis-speaking women — and the Bugis nobility whose standing depended in part on display of elaborately prepared food (cookies in particular) — have sought skill in cooking. It has been argued that competition among aristocrats, the wealthy and chefs played an important role in producing *haute cuisine* in China (Anderson 1988; Mennell 1996; Mennell *et al.* 1992): the Bugis case would seem to fit this model. In France as well as China, *haute cuisine* is largely the province of males (though in recent years it has become less gendered), in spite of women being generally responsible for ordinary, household cooking. Among the Bugis, men are not permitted to set foot in the kitchen, even during preparations for a *pesta*; the exceptions to this rule are the *bissu* (Ind. *waria*, male transvestites), who are said to be the best cooks of all.

Self-interest in cooking no doubt exists and keeping recipes secret or transmitting misleading information about them are probably common in some societies; in the Bugis case, however, there appears to have been little opportunity or motivation for such informational editing. Cooking knowledge is transmitted by direct observation. One learns to cook by watching and assisting one's elders, who are usually one's close relatives. The problems of cooking are, however, similar in one major respect to those faced by our distant ancestors: they involve a brief timeframe. As the proverb has it, 'the proof is in the eating'; that is, validation or corrective feedback, as the case may be, occurs shortly after a dish has been cooked. As for evolved mechanisms, one could argue that the sense of taste is indeed such a specialized mechanism: we seem to automatically become experts in the cuisine with which we are raised, and errors or other departures are instantly apparent to us.

Infant/child nutrition

Is child nutrition a matter of competition? Though no specific data were collected on this point, it seems reasonable to assume that in the study villages — and probably everywhere else — mothers take pride in the health and vitality of their offspring, and where there is pride there must be an element of competition. Women in the study sites regularly sought expert advice on infant and child health from the *dukun* and the staff of the community health centres and posts, showing their strong concern for the well-being of their children. However, whatever competition may or may not have existed concerning number of healthy children, it is not comparable to the formalized competition of a *pesta*, with its elaborately prepared dishes on display as a claim to and legitimation of relative social standing.

Does maternal self-interest bias child-feeding knowledge and practice? From the perspective of evolutionary biology, this is possible: a mother might enhance her own ability to have additional children by scanting those already born, or she might increase her completed fertility rate by favouring the child who would benefit more from a given amount of food at the expense of one who would benefit less (e.g. an ill child or a very young one could benefit more from, say, an egg, then one who was healthy or considerably older). In the present case, none of the feeding practices observed or indigenous knowledge provided by key informants and focus groups suggests any such self-interested biases. Indeed, though this possibility was not a research focus, it seems more likely that any food saved by the mother would be consumed neither by her nor by her children but by the adult males. Women learn infant and child-feeding practices from their own mothers; new mothers are given explicit feeding advice by the *sanro*, and today, by health post personnel as well.

Are child-feeding problems comparable to the problems faced by our ancestors? In a sense they are simply the same problems — infants and children must be fed — but this fact is misleading. If human intelligence evolved primarily as a response to the problems of social life, the relevant question is whether infant/child-feeding problems today are in any way comparable to the social problems faced by our ancestors. As with the case of Bugis cuisine, the dimension that can be readily compared with domains involving social life problems is that of timeframe for effective feedback. Unlike cooking, for which the timeframe is indeed comparable in being brief, the timeframe for feedback from child feeding is slow and complex, making error correction very difficult. Compared with the immediacy of determining whether or not a cooked dish tastes good, the effects of feeding a child fish or fruit a bit more often are very unclear and may not be manifest for months or years (while the danger of small bones choking a child can, alas, be learned quickly indeed!). Child health is determined by many factors, and children do not react uniformly to varying diets, so that corrective feedback with regard to nutrition is unreliable.

Have we evolved specialized mechanisms pertaining to infant feeding? It is instructive to compare the strength of our interest in the sexual and status-related activities of high-ranking members of our local group (that is, our tendency to gossip) with the strength of our interest in a child's diet; the former is so great that soap operas can transfix us, but there are no television shows showing the drama of permutations of diet on an infant's nutritional health! Similarly, we readily recall every instance in which we have aided another, especially when the aid may not have been fully reciprocated, but the details of what a child once ate, years past, are only vaguely recalled. We have apparently never been selected for the ability to recall precisely how much of what food was fed to which child when and then to track this information against the child's future growth and health. Why not?

Why have we not evolved some kind of specialized infant/child nutritional 'mental organ' or 'nutritional intelligence module' to keep track of child (and maternal) feeding and nutritional health? Here, one can only speculate. Perhaps existing adaptations, discussed earlier in terms of our evolved taste preferences, have generally been adequate to ensure proper diet for both mothers and children. Perhaps having much choice in what to feed a child is a relatively new phenomenon. Perhaps the selection pressure for a nutritional health module has long been there, but there is no substrate which could develop into such an organ (because something would be adaptive is hardly a guarantee that it will evolve, after all). Perhaps it is rare for child-feeding practices to have a strong impact on total reproductive success. Perhaps systems of food prohibitions and balance theories are relatively new in the history of our species, and there has not been sufficient time for us to evolve specialized abilities to counter them. Unfortunately, we have no data with which to evaluate these possibilities. The infant-feeding practice that seems most likely to have affected fitness is that of colostrum denial, but it is not clear that it has ever been sufficiently widespread, or that it has been practised through enough generations, for it to have produced a countering evolutionary response.

We now are in a position to answer the question of why the indigenous knowledge domain of cuisine seems so much more effective than the indigenous knowledge domain of child feeding/nutrition in the two study villages (and, presumably, in other Bugis communities as well). Cooking, especially in this particular society, turns out to be a good fit for our social intelligence: skill in cuisine has, at least historically, been strongly connected with social competition; our sense of taste is an evolved mechanism providing fairly unambiguous feedback about the effectiveness of our culinary efforts; and this feedback is in the short timeframe to which our intelligence appears to be well adapted. For child feeding/nutrition, social competition was not visible and, if present, is certainly muted; we appear to lack any specialized evolved mechanisms to permit us to correlate the details of past feeding practices with the health and growth of our children; the feedback that we do receive occurs in a lengthy timeframe and is highly ambiguous, given the many factors that influence infant/ child health. It is thus not surprising that Bugis indigenous knowledge of cooking is more effective than indigenous knowledge of child feeding and nutrition.

THE KNOWLEDGE BASES OF HUMAN SOCIAL INSTITUTIONS

This serendipitous comparison of indigenous knowledge of cuisine versus nutrition has produced a general, testable set of hypotheses: accurate and extensive domains of indigenous knowledge are most likely to develop when they are (or have been) foci of social competition, when they can benefit from specialized evolved mechanisms, and when the timeframe of corrective feed-back from physical or social reality is rather short. Study of indigenous knowl-edge domains of many kinds and in many societies will be needed to evaluate these hypotheses and answer a range of questions. For example, do the hypotheses apply to agricultural knowledge, or to the raising of livestock? How short is 'short'? What is the impact of 'borrowing' knowledge and personnel from neighbouring societies?

Because in this chapter we are concerned with the origins of human social institutions, let us for the moment assume that future research will validate the hypotheses we have here developed. If so, then we can provisionally conclude that elaborate social organization and institutions would have arisen quite readily among our ancestors, because our social brains, specialized as they are in solving social problems, would have been able to generate the organizational knowledge needed. The more competition there was among individuals and groups involved in creating this knowledge, the more quickly it would have been produced and elaborated and the more effective it would have become. However, size and complexity of organization would have affected the speed and accuracy of learning from the consequences of actions: as a society grows larger, it seems likely that the time between the making of social organization decisions by leaders and the learning of the consequences of those decisions will grow longer.

The growth of technological knowledge, because our brains are not partic-ularly well adapted to its production, seems especially likely to have been dependent on social competition. The hypothesis that the societies which developed effective technologies early were those that created competition among artisans suggests itself. For example, societies fostering competition among builders would develop a knowledge basis for construction more quickly than societies in which there was no such competition.

What of the situation in which there was strong social competition in a domain of knowledge that perforce offers little possibility of corrective feed-back from an external reality, e.g. the realms of religion and philosophy? Here, too, wherever a situation of competition was created, we would expect an extensive generation of knowledge to result. Such knowledge domains would have and still do produce, at times, much beauty and perhaps insight into the human condition; however, because in no timeframe could there be corrective feedback from external reality, effective knowledge would not necessarily accumulate, and political processes would have been the main determinants of the relative standing of its producers. Even today, one could argue that the distinction between the humanities and the sciences is simply that, while for both political processes play major roles in the evaluation of the relative worth of intellectual contributions, the knowledge produced in the sciences is influ-enced by systematically sought corrective feedback from external reality, while in the humanities there is no such dilution of political processes; instead, contending circles strive to make their own leaders and criteria for excellence of knowledge dominant (e.g. the debate over which literary works are to be included in the 'canon').

Students of human intelligence would do well to consider the archaeological record, along with the various domains of indigenous knowledge of extant societies, as natural experiments. To study cultural knowledge domains and the social processes whereby they may be created, maintained, edited, revised and deleted is to study the intelligence of our species *in situ*.

Finally, this analysis does yield at least one piece of practical advice concerning food: if you seek fine cuisine, go where there is a long history of competition among cooks.

Note. This research was sponsored by ISLE (Island, Sustainability, Livelihood, Equity), a 1996–2000 Canadian International Development Agency project administered by Dalhousie University and involving, in addition to that university, the Nova Scotia Agricultural College, the University of Prince Edward Island, the University of the West Indies, the University of the Philippines (Visayas), and Indonesia's Hasanuddin University. The authors wish to thank CIDA as well as ISLE's patient and encouraging director, Dr Gary Newkirk, and its administrator, the kind and efficient Ms Pauline Peters. Thanks are also due to the staff of Dalhousie's Lester Pearson Institute, Ms Becky Field in particular. The data were collected in connection with an ISLE-sponsored multidisciplinary course on 'Island Food Systems' which was organized and co-ordinated by Professor Claude Caldwell of NSAC, who was always generous with his support. Thanks are also owed for their warm helpfulness to the Bupati of Boné, Andi Muhammad Amir, and to the secretary to the Bupati, Dr H.A. Mappamadeng Dewang, as well as to Mr Murtir Jeddawi, Chief of the Boné District Planning and Development Board. We also wish to thank the following individuals for their helpfulness and generosity: Mr Muchlis A. Rasyid and Ms Taswina A. Muchlis, head of the Tanete Riattang Timur Subdistrict and head of its Women's Movement, respectively; Mr A. Bachtiar and Ms A. Bachtiar, head of Panyula Village (of the Tanete Riattang Timur Subdistrict) and head of the Panyula Village Women's Movement, respectively; Mr A. Lantara, head of Amali Subdistrict, and Ms A. Lantara, head of its Women's Movement; and Dr M.Y. Sara, director of the Taretta Village Health Centre. Finally, our deepest appreciation to the people of the villages of Taretta and Panyula, who were very kind to and open with the inquisitive strangers in their midst. Barkow is responsible for this paper's theoretical analysis. Correspondence concerning the paper should be addressed to him (*j.h.barkow@dal.ca*). Nurpudji was the principal investigator for the project and she and Veni Hadu were responsible for collecting data on nutritional status and dietary intake; they and Ramli were responsible for the nutritional analysis. Sani Silwana, Yahya, Djunaidi and Barkow were responsible for collecting foodways material. Faisal Attamimi and Elly Ishak were responsible for the food safety analysis. Barkow wishes to thank Marta Mahler for the suggestion that the danger to small children of choking on fishbones may outweigh the benefits of additional protein, and Susan Millar for helpful e-mail discussion of cooking and competition in Bugis society. Any errors are of course his.

REFERENCES

ANDERSON, E.N. 1988: *The Food of China* (New Haven, CT, Yale University Press).

ANTWEILER, C. 1998: Local knowledge and local knowing — an anthropological analysis of contested 'cultural products' in the context of development. *Anthropos* 93, 469–94.

AUNGER, R. 1994a: Are food avoidances maladaptive in the Ituri Forest of Zaïre? *Journal of Anthropological Research* 50, 277–310.

AUNGER, R. 1994b: Sources of variation in ethnographic interview data — food avoidances in the Ituri Forest, Zaïre. *Ethnology* 33, 65–99.

BARKOW, J.H. 1989a: *Darwin, Sex, and Status: Biological Approaches to Mind and Culture* (Toronto, University of Toronto Press).

BARKOW, J.H. 1989b: Overview. *Ethology and Sociobiology* 10, 1–10.

BARKOW, J.H. 1992: Beneath new culture is old psychology. In Barkow, J.H., Cosmides, L. & Tooby, J. (eds.), *The Adapted Mind: Evolutionary Psychology and the Generation of Culture* (New York, Oxford University Press), 626–37.

BARKOW, J.H. & HALLETT, A.L. 1989: The denial of colostrum. In Barkow, J.H. (ed.), *Darwin, Sex and Status: Biological Approaches to Mind and Culture* (Toronto, University of Toronto Press), 301–9.

BILLING, J. & SHERMAN, P.W. 1998: Antimicrobial functions of spices: why some like it hot. *Quarterly Review of Biology* 73, 3–49.

BLACKMORE, S. 1999: *The Meme Machine* (New York, Oxford University Press).

BLUM, J.W., HADORN, J.-P.S. & SCHUEP, W. 1997: Delaying colostrum intake by one day impairs plasma lipid, essential fatty acid, carotene, retinol and tocopherol status in neonatal calves. *Journal of Nutrition* 127, 2024–9.

BOGIN, B. 1997: The evolution of human nutrition. In Romanucci-Ross, L., Moerman, D.E. & Tancredi, L.R. (eds.), *The Anthropology of Medicine: From Culture to Method*, 3rd edn (Westport, CT, Bergin & Garvey), 96–142.

BOYD, R. & RICHERSON, P.J. 1985: *Culture and the Evolutionary Process* (Chicago, University of Chicago Press).

BOYER, P. 1998: Cognitive tracks of cultural inheritance: how evolved intuitive ontology governs cultural transmission. *American Anthropologist* 100, 876–89.

BRACE, L. 1996: Modern human origins and the dynamics of regional continuity. In Akazawa, T. & Szathmary, E. (eds.), *Prehistoric Mongoloid Dispersals* (New York, Oxford University Press), 81–112.

BURRIN, D.G., DAVIS, T.A., EBNER, S., SCHOKNECHT, P.A., FIOROTTO, M.L. & REEDS, P.J. 1997: Colostrum enhances the nutritional stimulation of vital organ protein synthesis in neonatal pigs. *Journal of Nutrition* 127, 1284–9.

BYRNE, R. & WHITEN, A. (eds.) 1988: *Machiavellian Intelligence: Social Expertise and the Evolution of Intellect in Monkeys, Apes, and Humans* (Oxford, Oxford University Press).

COSMIDES, L. & TOOBY, J. 1992: Cognitive adaptations for social exchange. In Barkow, J.H., Cosmides, L. & Tooby, J. (eds.), *The Adapted Mind: Evolutionary Psychology and the Generation of Culture* (New York, Oxford University Press), 163–228.

CRONK, L. 1991: Wealth, status, and reproductive success among the Mukogodo of Kenya. *American Anthropologist* 93, 345–60.

DAWKINS, R. 1976: *The Selfish Gene* (Oxford, Oxford University Press).

DETTWYLER, K.A. 1994: *Dancing Skeletons: Life and Death in West Africa* (Prospect Heights, IL, Waveland).

EATON, S.B., SHOSTAK, M. & KONNER, M. 1988: *The Paleolithic Prescription: a Program of Diet and Exercise and a Design for Living* (New York, Harper & Row).

ELLEN, R.F. & HARRIS, H. 1997: *Indigenous Environmental Knowledge in Scientific and Developmental Literature* (Canterbury, University of Kent at Canterbury).

FERRO-LUZZI, G.E. 1973: Food avoidances at puberty and during menstruation in Tamiland. *Ecology of Food and Nutrition* 2, 165–72.

GRENIER, L. 1998: *Working with Indigenous Knowledge: a Guide for Researchers* (Ottawa, IDRC).

HADORN, U., HAMMON, H., BRUCKMAIER, R.M. & BLUM, J.W. 1997: Delaying colostrum intake by one day has important effects on metabolic traits and on gastrointestinal and metabolic hormones in neonatal calves. *Journal of Nutrition* 127, 2011–23.

HARRIS, M. 1987: Foodways: historical overview and theoretical prolegomenon. In Harris, M. & Ross, E.B. (eds.), *Food and Evolution: Toward a Theory of Human Food Habits* (Philadelphia, Temple University Press), 57–90.

HAWKEN, C.M. 1999: *Colostrum: Amazing Immune System Enhancer* (Pleasant Grove, UT, Woodland Publishing).

HULL, V.J. 1986: Dietary taboos in Java: myths, mysteries, and methodology. In Manderson, L. (ed.), *Shared Wealth and Symbol: Food, Culture and Society in Oceania and Southeast Asia* (Cambridge, Cambridge University Press), 237–58.

HUMPHREY, N.K. 1976: The social function of intellect. In Bateson, P.P.G. & Hinde, R.A. (eds.), *Growing Points in Ethology* (Cambridge, Cambridge University Press), 303–18.

HUMPHREY, N.K. 1983: *Consciousness Regained: Chapters in the Development of Mind* (Oxford, Oxford University Press).

KATONA-APTE, J. 1977: The sociocultural aspects of food avoidances in a low-income population in Tamiland, South India. *Journal of Tropical Pediatrics and Environmental Child Health* 23, 89–90.

KATZ, S.H., HEIDEGER, M.L. & VALLEROY, L.A. 1974: Traditional maize-processing techniques in the new world. *Science* 184, 765–73.

KIKAFUNDA, J.K., WALKER, A.F., COLLETT, D. & TUMWINE, J.K. 1998: Risk factors for early childhood malnutrition in Uganda. *Pediatrics* 102, E451–E458.

LADERMAN, C. 1984: Food ideology and eating behavior: contributions from Malay Studies. *Social Science and Medicine* 19, 547–59.

LEY, B.M. 1997: *Colostrum: Nature's Gift to the Immune System* (El Cajon, CA, Nutri-Books).

LUMSDEN, C.J. & WILSON, E.O. 1981: *Genes, Mind and Culture* (Cambridge, MA., Oxford University Press).

MANDERSON, L. 1986: Food classification and restriction in peninsular Malaysia: nature, culture, hot and cold? In Manderson, L. (ed.), *Shared Wealth and Symbol: Food, Culture and Society in Oceania and Southeast Asia* (Cambridge, Cambridge University Press), 127–43.

MENNELL, S. 1996: *All Manners of Food: Eating and Taste in England and France from the Middle Ages to the Present*, 2nd edn (Urbana, IL, University of Illinois Press).

MENNELL, S., MURCOTT, A. & VAN OTTERLOO, A.H. 1992: *The Sociology of Food: Eating, Diet and Culture* (London, Sage).

MILLAR, S.B. 1989: *Bugis Weddings. Rituals of Social Location in Modern Indonesia* (Berkeley, CA, Center for South and Southeast Asia Studies/University of California at Berkeley).

MILLER, G. 2000: *The Mating Mind: How Sexual Choice Shaped Human Nature* (New York, Doubleday).

NYGREN, A. 1999: Local knowledge in the environment-development discourse: from dichotomies to situated knowledge. *Critique of Anthropology* 19, 267–88.

PELRAS, C. 1996: *The Bugis* (Oxford, Blackwell).

RHOADES, R. & BEBBINGTON, A. 1995: Farmers who experiment: an untapped resource for agricultural research and development. In Warren, D.M., Slikkerveer, L.J. & Brokensha, D. (eds.), *The Cultural Dimension of Development: Indigenous Knowledge Systems* (London, Intermediate Technology Systems), 296–307.

ROSENBERG, E.M. 1980: Demographic effects of sex-differential nutrition. In Jerome, N.W., Kandel, R.F. & Pelto, G.H. (eds.), *Nutritional Anthropology: Contemporary Approaches to Diet and Culture* (Pleasantville, NY, Redgrave), 181–203.

ROZIN, P. 1987: Psychobiological perspectives on food preferences and avoidances. In Harris, M. & Ross, E.B. (eds.), *Food and Evolution: Toward a Theory of Human Food Habits* (Philadelphia, Temple University Press), 181–205.

SEMALI, L.M. & KINCHELOE, J.L. (eds.) 1999: *What is Indigenous Knowledge? Voices From the Academy* (New York, Falmer Press).

SILLITOE, P. 1998: The development of indigenous knowledge: a new applied anthropology. *Current Anthropology* 19, 223–52.

SPERBER, D. 1996: *Explaining Culture: a Naturalistic Approach* (Oxford, Blackwell).

VAN ESTERIK, P. 1989: *Beyond the Breast–Bottle Controversy* (New Brunswick, NJ, Rutgers University Press).

WARREN, D.M., SLIKKERVEER, L.J. & BROKENSHA, D. (eds.) 1995: *The Cultural Dimension of Development: Indigenous Knowledge Systems* (London, Intermediate Technology Publications).

WFOOD2. 1994: *World Food Program Dietary Assessment System* (Berkeley, University of California).

WHITEN, A. & BYRNE, R.W. 1997: Preface. In Whiten, A. & Byrne, R.W. (eds.), *Machiavellian Intelligence II: Extensions and Evaluations* (Cambridge, Cambridge University Press), 1–23.

WHO 1998: *Complementary Feeding of Young Children in Developing Countries: a Review of Current Scientific Knowledge* (Geneva, World Health Organization).

WILSON, C.S. 1973: Food taboos of childbirth: the Malay example. *Ecology of Food and Nutrition* 2, 267–74.

WILSON, C.S. 1986: Social and nutritional context of 'ethnic' foods: Malay examples. In Manderson, L. (ed.), *Shared Wealth and Symbol: Food, Culture and Society in Oceania and Southeast Asia* (Cambridge, Cambridge University Press), 259–72.

WOLFF, R.J. 1965: Meanings of food. *Tropical and Geographical Medicine* 1, 45–61.

WRANGHAM, R.W., JONES, J.H., LADEN, G., PILBEAM, D. & CONKLIN-BRITTAIN, N. 1999: The raw and the stolen: cooking and the ecology of human origins. *Current Anthropology* 40, 567–94.

How and Why Did Fairness
Norms Evolve?

KEN BINMORE

> What is just . . . is what is proportional.
>
> Aristotle, *Nicomachean Ethics*

WHAT IS FAIR?

WHEN A DISH IN SHORT SUPPLY is shared at a polite dinner party, there is seldom any verbal dispute. If things go well, the dish is divided without any discussion or intervention by the host. When questioned, everybody will agree that each person should take his or her fair share. But how do we know what is fair?

This is not a simple question. What is judged to be fair according to our current standards of morality depends on a complex combination of contingent circumstances — like who is fat and who dislikes cheese. Moreover, if we observe what actually happens, rather than what people say should happen, we will find that it also depends on how each person at the table fits into the social pecking order. Woe betide the poor relative sitting at the table on sufferance in the nineteenthcentury who helped himself to an over-generous portion of his favourite dish!

Numerous scholars have tried to make sense of the calculations that people must implicitly have made when they co-ordinate on an outcome that they afterwards describe as fair. It surely can be no accident that the consensus is firmly in favour of some type of do-as-you-would-be-done-by principle. Moralists down the ages have offered numerous arguments that seek to explain why it is morally imperative that each person should follow such a golden rule. But none of these traditional arguments is founded on anything solid. I think we become suckered into taking them seriously because we are too ready to confuse a fairly accurate description of *what* we do in certain circumstances with an explanation of *why* we do it.

Rather than resorting to metaphysical speculation, I think that the first step on the road to understanding the human thirst for justice lies in the recognition that variants of the do-as-you-would-be-done-by principle are *already* firmly entrenched among the instincts and customs that regulate our lives. The relevant norms do not survive because we consciously cherish them. On the

Proceedings of the British Academy, **110**, 149–70, © The British Academy 2001.

contrary, I think that most of our habituated behaviour is acquired using processes that operate below the level to which our conscious minds have easy access. Like monkeys, we are programmed to imitate the behaviour of our more successful neighbours. If those in thrall to a particular habit or custom are perceived as being winners, then their habituated behaviour will be copied, without any need for anyone to understand *why* the habituated behaviour works well in the current social environment.

A fairness norm may be a do-as-you-would-be-done-by principle, but many such principles can be formulated. Which of these should we study? To my knowledge, only one principle has been proposed that adequately responds to objections such as: Don't do unto others as you would have them do unto you — they may have different tastes from yours. This chapter will need to refer to both Rawls (1972) and Harsanyi (1977) in studying this fairness principle, but the terminology will be that of Rawls' *Theory of Justice*. Rawls proposes the *original position* as a hypothetical standpoint to be used in making judgements about how a just society should be organized. Each citizen is asked to envisage the social contract to which he or she would agree *if* his or her current role in society were concealed behind a *veil of ignorance*. In considering the social contract on which to agree under such hypothetical circumstances, each person will pay close attention to the plight of those who end up at the bottom of the social heap. Devil take the hindmost is not such an attractive principle when you yourself may be at the back of the pack.

I think that the reason most people find the device of the original position intuitively attractive as a fairness criterion has nothing to do with the Kantian arguments offered by Harsanyi and Rawls. I believe that its appeal lies in the fact that we recognize it as a stylized version of a principle that we already unconsciously apply every day when interacting with our peers. From such a perspective, fairness is interpreted entirely in naturalistic terms. The original position is merely a device that has been washed up on the beach along with the human race by the forces of biological and social evolution. If we can figure out precisely how we use it at present to avoid inefficient disputes over small matters, perhaps we will also be able to use it to achieve stable political compromises over large-scale issues. The defence for such a proposal is entirely pragmatic. Here is a tool supplied by Nature. Let us use it to improve our lives, just as we use whatever tools we find in our toolbox when making repairs around the house. But we shall get nowhere in this enterprise if we refuse to be realistic about how the device of the original position functions in our daily life at present.

Psychological equity theory

Our capacity for objective introspection is notoriously limited. What we say about our beliefs and motivations is often absurdly at variance with our

behaviour. Experimental work is therefore necessary to discover how we actually split a surplus when we believe ourselves to be acting fairly.

Social psychologists who have conducted experiments on fairness have
been led to an empirically based law that resolves problems of social exchange
by equalizing the ratio of each person's gain to his or her worth (Furby 1986;
Mellers 1982; Mellers and Baron 1993; Walster *et al.* 1978). People who are
deemed worthy therefore get more of the gravy than others. As in Wilson
(1993), this theory is usually referred to as 'modern equity theory', although it
originates with Aristotle's *Nicomachean Ethics* and has been little developed
since it was introduced to social psychologists by Homans (1961) and Adams
(1963, 1965) more than thirty years ago. Selten (1978) provides an account of
the theory which is easily accessible to economists.

The psychological theory of equity requires that a surplus be shared in proportion to each person's worthiness. Written as an equation:

$$\frac{g_A}{w_A} = \frac{g_E}{w_E} \tag{1}$$

where g_A and g_E are the respective gains to Adam and Eve, and w_A and w_E quan-
tify how worthy they are. But how are gains to be measured? Where is the zero
to be located on whatever scale is chosen? How is worthiness to be construed?
Is it to be measured in terms of social status, merit, effort, need or what? My
understanding of the pyschological literature is that the answers to these questions depend on the context. But what is the rule that maps a context onto the
relevant scales for measuring gain and worthiness?

To answer such questions, one needs a background theory to suggest critical experiments. I believe that such a theory can be constructed by asking how
the apparatus of the original position proposed by Harsanyi and Rawls may
have evolved from prehistoric food-sharing agreements between members of
the same family. In seeking to construct such a theory in the following pages,
one needs to make hypotheses about the social contracts that held sway among
human foraging bands in prehistory. The second half of this chapter takes up
the main purpose of the paper, which is to comment on the claim that modern
hunter-gatherer societies provide a suitable model for their prehistoric coun-
terparts. The rest of the chapter briefly relates this commentary to my adaptation of the theories of Harsanyi and Rawls.

Natural duty?

Rawls (1972) invented the device of the original position to provide a properly
argued alternative to utilitarianism. Harsanyi (1977) appealed to precisely the
same device when defending utilitarianism. I support Harsanyi in this dispute,

since Rawls succeeds in evading a utilitarian conclusion only by throwing orthodox decision theory overboard. However, I think that Rawls' intuitive grasp of the type of outcome to which one is led by applying the original position under realistic conditions is much sounder. Rawls advocates redistributing worldly goods according to the maximin criterion, which demands that we give priority to ensuring that the worst-off members of society gets as much as possible.

Figure 1(a) compares Rawls' maximin outcome R with Harsanyi's utilitarian outcome H. In this diagram, Adam and Eve are the two members of a society inhabiting the Garden of Eden. A social contract is modelled as a pair $x = (x_A, x_E)$ of utilities. The set X contains the social contracts that are feasible. Figure 1(b) shows three bargaining solutions from co-operative game theory: the Nash bargaining solution n; a weighted utilitarian solution h; and the proportional or egalitarian bargaining solution r. The point ξ represents Adam and Eve's current *status quo*. Our focus for the moment is on the third of these.

Two important features of the egalitarian bargaining solution should be noted. The first is that r can be identified with the result of applying the psychological equity law if Adam and Eve's respective gains are taken to be $g_A = r_A - \xi_A$ and $g_E = r_E - \xi_E$, and their worthiness coefficients are chosen so that w_E/w_A is the slope of the line joining ξ and r. The second point is that r is also the result of applying the maximin criterion after correcting x_A and x_E to $(x_A - \xi_A)/w_A$ and $(x_E - \xi_E)/w_E$. Such a correction corresponds to relocating the zeros and units on Adam and Eve's utility scales in order to ensure that our standard of measurement matches the manner in which interpersonal comparisons of welfare are made in the society under study.

I think Rawls' attempt to derive the maximin criterion from an analysis of how Adam and Eve will bargain behind the veil of ignorance goes awry at two

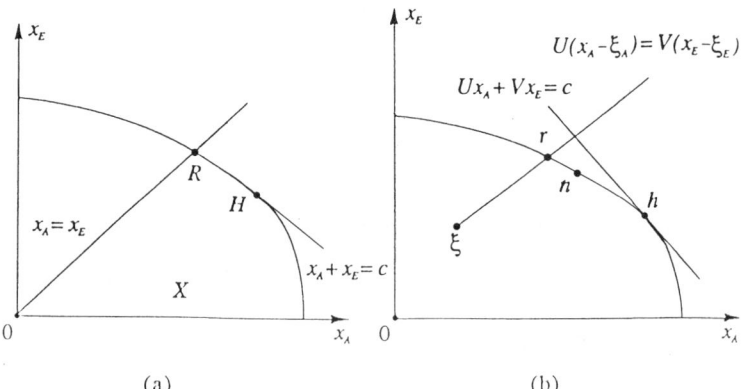

                                  (a)   (b)

Figure 1. Rawls and equity theory.

points. He should not have adopted the iconoclastic expedient of denying orthodox decision theory, and he should not have joined with Harsanyi in assuming that Adam and Eve are *committed* to the hypothetical deal reached in the original position. Rawls (1972: 115) says that we have a 'fundamental natural duty . . . to comply with just institutions', but I think that he and Harsanyi are really just indulging in some wishful thinking. It would certainly make life more pleasant if we instinctively rated the call of justice above our own selfish concerns, but the evidence for such a claim is not very favourable.

The commitment problem arises in its starkest form in the study of the Prisoners' Dilemma of Figure 2(a). If Adam and Eve discuss how they should play this game, whether behind a veil of ignorance or not, they are likely to agree that both should play *dove*. Each will then receive a payoff of 2. If they are committed to the agreement, this is the end of the story. But if they are not committed, then they have the opportunity to cheat on the agreement when the time comes to play. Since cheating on the deal by playing *hawk* is optimal for each player whatever strategy the other chooses, the result will be that both play *hawk*. Each then receives a payoff of 0.

When Adam and Eve both choose *hawk* in the Prisoners' Dilemma, each is using a strategy that is an optimal reply to the strategy choice of the other. Game theorists register that a pair of strategies has this property by calling it a Nash equilibrium. If an authoritative book on game theory records the rational solution to a game, it must be a Nash equilibrium — otherwise it would be rational for at least one player to deviate from the book's recommendation.

Various attempts to escape the conclusion that rational play calls for both players to cheat in the one-shot Prisoners' Dilemma have been proposed which

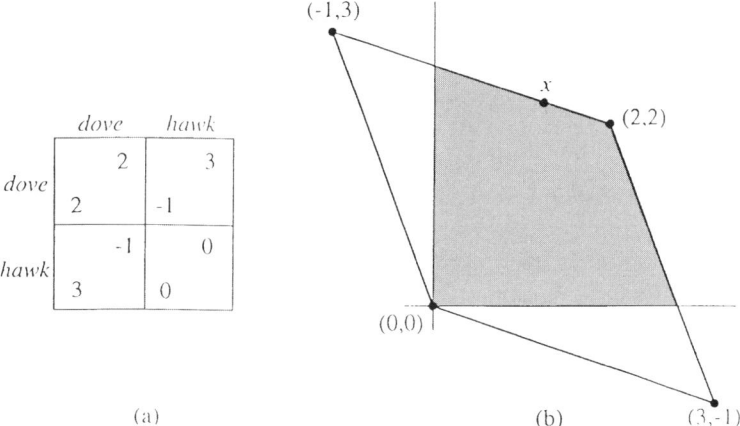

Figure 2. Prisoners' Dilemma.

postulate that Nature has equipped us with *internal* commitment mechanisms whose engagement can be convincingly transmitted to an opponent (Binmore 1994). But where is the evidence that such internal mechanisms exist? Why would they be evolutionarily stable? Since no adequate answers are on offer, game theorists restrict their attention to *external* commitment mechanisms. For example, if Adam and Eve sign a legal contract under modern circumstances to play *dove* in the Prisoners' Dilemma, then each will regard themselves as committed to the agreement, since any breach of the contract will be punished by our judicial system. Other external enforcement agencies have operated in other places and at other times. Fear of ostracism by one's peer group is a particularly effective form of disciplining agreements.

In postulating an evolutionary history for the device of the original position, it is therefore important that we take a view on the extent to which an external source of authority for policing agreements was available in the relevant period of prehistory. If an external enforcement agency were available in the form of a dominant leader or strong peer pressure, then Harsanyi's (1977) analysis suggests that our fairness norms would be utilitarian in character. However, when a similar analysis is applied to the case when no external enforcement agency at all exists, Binmore (1994, 1998) is led to fairness norms that implement the egalitarian bargaining solution. The latter conclusion is more in line with modern experiments on fairness, but one would have to dismiss this as a coincidence if a case could not be made for the claim that prehistoric hunter-gatherer bands operated so anarchic a social contract that agreements between pairs of individuals were viable only if the pair themselves were able to police the agreement without help from other members of the band.

Reciprocity

The one-shot Prisoners' Dilemma is very misleading if used as a model of the human game of life. Since its only Nash equilibrium requires each player to cheat on any co-operative agreement, we would not have evolved as social animals if it were our game of life. As explained first by Hume in his *Treatise of Human Nature* in 1739, the mechanism that sustains human co-operation is *reciprocity*. But Adam cannot threaten not to scratch Eve's back if she won't scratch his, without presupposing that they have an ongoing relationship to nourish. To model such self-policing, long-term relationships, we need to study the Nash equilibria of games that are to be *repeated* an indefinite number of times. If the players are sufficiently forward-looking that future payoffs seem nearly as good as current payoffs, they will be reluctant to cheat on their partners today for fear of losing the fruits of co-operation tomorrow.

Trivers (1971) introduced this idea into biology under the name of *reciprocal altruism*. Axelrod (1984) popularized the notion further by explaining why it is a Nash equilibrium in the indefinitely repeated Prisoners' Dilemma for each player to use the strategy TIT-FOR-TAT. Since the resulting outcome is that each player receives a payoff of 2 each time the Prisoners' Dilemma is repeated, one learns that rational co-operation is possible without any need to call upon the services of an external enforcement agency. However, the fuss about TIT-FOR-TAT obscures the fact that the problem in studying an indefinitely repeated game is not *whether* co-operative equilibria exists, but *which* of the many co-operative equilibria should be selected.

In the early 1950s, before Trivers or Axelrod, several game theorists independently discovered the *folk theorem* that characterizes the whole set of Nash equilibrium outcomes of an indefinitely repeated game (Aumann & Maschler 1995). For example, the shaded region of Figure 2(b) is the set of all per-game payoff pairs that can be supported as equilibria by sufficiently forward-looking players in the indefinitely repeated Prisoners' Dilemma. The *equilibrium selection problem* in such a game consists of predicting which of these outcomes will actually be observed. The symmetry of the Prisoners' Dilemma makes the symmetric outcome (2,2) focal but real-life games are seldom symmetrical. To create a co-operative species, Nature therefore had to find a way of allowing equilibrium selection devices to evolve. I believe that fairness is one of Nature's solutions to this problem.

If one accepts that fairness norms evolved to co-ordinate behaviour on an equilibrium in a repeated game of life in the absence of any external enforcement agency, then one must also accept that the procedure required to implement the fairness norm must be as self-policing as the equilibrium it is designed to select. Far from postulating a natural duty to be just, I therefore assume that people will cheat on the judicial procedure whenever they can. The only procedures that are viable are therefore those that provide nobody with a motive to cheat. As observed in the previous section, adopting this principle requires that the approaches of both Harsanyi and Rawls be very substantially modified. Rather than being led to the utilitarian outcome that results if one applies orthodox decision theory with external enforcement, one is led instead to an egalitarian outcome *r* as illustrated in Figure 1(b).

However, a major problem remains. The worthiness coefficients w_A and w_E are undetermined in our specification of the egalitarian bargaining solution. But we need to know what they are if we are to apply the egalitarian solution to the problem of selecting an equilibrium in a game like the indefinitely repeated Prisoners' Dilemma.

Interpersonal comparison of utility

The laboratory experiments that led psychologists to formulate their equity law suggest that modern fairness norms are egalitarian rather than utilitarian, but further experimentation has been hindered by lack of a background theory able to make predictions about how the worthiness coefficients w_A and w_E should be anticipated to vary with the context. So what does my theory have to say on this subject?

I argue that the food-sharing agreements with which human co-operation presumably began originated within the family. Since we share genes with our kin, it would be surprising if we were not biologically programmed to write their welfare into our utility functions according to their degree of relationship to us. For example, according to Hamilton's (1963, 1964) rule, if Eve is Adam's full cousin, then he should care for her one-eighth as much as he cares for himself. The reason is that the probability that her body is playing host to any specific gene in his body is 1/8. My guess is therefore that we are biologically hardwired to assess the probable degree of relationship to those we encounter within the family circle, and to use this as a standard for making interpersonal comparisons when comparing their lot with our own.

But the interesting case consists of our fairness transactions with strangers. I believe that the fairness algorithm itself is biologically hardwired, but that its adaptation for use with strangers must have been contrived by *cultural* evolution. We learned to adopt strangers into our clans by treating them as relatives. But the degree of relationship attributed to such adopted strangers must have been socially determined. However, if the worthiness of someone outside the family circle is a social convention, then it need not be constant as the context varies. Nor need it be invulnerable to change over time.

The latter consideration is particularly important, since it allows predictions to be made about how worthiness coefficients will adjust over time in a fixed context. In Binmore (1998), I argue that one must expect social evolution to change the way in which people perceive the worthiness of others until the egalitarian bargaining solution r of Figure 1(b) coincides with the Nash bargaining solution n. In principle, one can then predict the relative size of w_A and w_E under ideal conditions. First locate the Nash bargaining solution for the feasible set X with *status quo* ξ. The ratio w_A/w_E is then the slope of the line joining ξ and n.

ANTHROPOLOGICAL EVIDENCE

In this section, I turn to the main business of this chapter, which is to assess the anthropological evidence for and against the assumptions of the theory briefly outlined in the foregoing pages. It first needs to be noted that the consensus is strong among anthropologists that uncontaminated hunter-gatherers, from Greenland eskimos to Kalahari bushmen, operated sharing-caring societies without bosses or social distinctions. The sharing of food, especially meat, is reported to be universal. (Bailey 1991; Damas 1972; Erdal & Whiten 1996;

Evans-Pritchard 1940; Gardner 1972; Hawkes *et al.* 1993; Helm 1972; Isaac 1978; Kaplan & Hill 1985; Knauft 1991; Lee 1979; Megarry 1995; Meggitt 1962; Riches 1982; Rogers 1972; Sahlins 1974; Tanaka 1980; Turnbull 1965. Usually, modern foraging societies are said to be egalitarian, but I prefer not to use this word in a sense that would include a utilitarian society.)

Why share food?

I follow the traditional line that attributes the evolutionary origins of the food-sharing phenomenon to the need for individuals to insure each other against privation. As Evans-Pritchard (1940: 85) explains:

> The habit of share and share alike is easily understandable in a community where everyone is likely to find himself in difficulties from time to time, for it is scarcity and not sufficiency that makes people generous, since everybody is then insured against hunger. He who is in need today receives help from him who may be in need tomorrow.

At least three criticisms of this explanation of food sharing need to be mentioned. The first is that prehistoric hominids are unlikely to have been provided with the 'Machiavellian intelligence' necessary to sustain such insurance contracts (Byrne & Whiten 1988). This piece of jargon expresses the familiar claim that to explain a piece of behaviour in terms of rational self-interest is to assert that it was carefully planned in advance by a coldly calculating intellect. Economists commonly disclaim such straw men by pointing out that a person riding a motorbike is implicitly solving a very difficult mathematical control problem, but nobody would think to deduce that Hells Angels must therefore be master mathematicians. As Evans-Pritchard explains, people in hunter-gatherer societies acquire the *habit* of sharing — and this habit survives because it co-ordinates behaviour on an equilibrium of the game of life without anyone even needing to be aware that a game is being played.

The second and third criticisms arise from differences between the situation envisaged by Evans-Pritchard and more recent reports of modern hunter-gatherer societies. The second criticism disputes the suggestion that hunter-gatherer societies commonly live on the edge of extinction. Sahlins (1974) observes to the contrary that modern hunter-gatherer societies have a relatively affluent lifestyle if one compares the amount of leisure they enjoy after meeting their needs with that of an agricultural labourer or a university professor. But we should not assume that the natural methods of birth control with which modern hunter-gatherers help to regulate their populations preceded the evolution of the food-sharing phenomenon. My guess is that population control is a relatively recent adaptation to the marginal territories currently occupied by hunter-gatherers. But, without controls of some kind, the iron law of Malthus would soon turn plenty into scarcity. How else does one explain the spread of hunter-gatherer societies over the whole world, even to the most inhospitable of environments? Even for modern hunter-gatherers, every year cannot be a fat year, and it is in the lean years that the invisible hand of evolution strikes down unfit groups. Nor can hunters rely on bringing home the bacon even in the fattest years, so that there will always be good reasons for sharing meat on a reciprocal basis.

The third criticism challenges Evans-Pritchard's appeal to reciprocity as an explanatory factor in food sharing. For example, Erdal and Whiten (1996) conclude that their survey of more than a hundred studies demonstrates that the sharing of food observed 'goes beyond the explanatory power of either kinship or reciprocation. Individuals do sometimes attempt to obtain a disproportionate share of resources or influence for themselves, but this is contained through vigilance and counter-dominant behaviour by their group members.' But what is the second sentence about if not a social contract in which everybody looks after everybody else because those who don't are punished by their fellows? It is true that the mechanism that supports the reciprocal arrangement is not one of the simple models of bilateral exchange that people usually have in mind when they refer to TIT-FOR-TAT. But the punishment strategies that support efficient equilibria in repeated games do not necessarily require that the player injured by a deviant is also the person who punishes the deviation. In the case of modern hunter-gatherer societies, the whole band combines to act as an external enforcement agency in punishing anyone who fails to co-operate in operating the scheme of mutual insurance by means of which it succeeds in surviving when times are bad.

However, although I think that Erdal and Whiten go astray in thinking that the data they survey cast doubt on theories that model hunter-gather societies in terms of rational self-interest, their survey makes it necessary to think twice about the important issue of how authority operates among modern hunter-gatherer societies as compared with the prehistoric societies in which our capacity for making fairness judgements presumably evolved.

Enforcement in foraging societies

Knauft (1991) argues that the evolution of authority in human societies can be seen in terms of a U-shaped curve, in which dominance-structured pre-human societies gave way to anarchic bands of human hunter-gatherers that were then replaced by the authoritarian herding and agricultural societies with which recorded history begins. As Erdal and Whiten (1996) document, the evidence is strong that leadership in modern hunter-gatherer societies lies only in influencing the consensus: 'But when a consensus has been reached, no-one has to follow it against their will — there is no enforcement mechanism.'

At first sight, the apparently anarchic structure of modern hunter-gatherer societies would seem to support my claim that external enforcement structures were indeed absent when the fairness algorithms I believe to be biologically determined were evolving. However, one has to be careful not to put the cart before the horse. One cannot argue that food sharing is the key to human sociality, and simultaneously proceed as though humans were already living in organized communities in the style of modern hunter-gatherers when the fairness norms governing the sharing of food evolved. Nor does the fact that modern hunter-gatherers operate social mechanisms that prevent potentially authoritarian leaders from becoming established imply that their societies do not enforce norms. On the contrary, the evidence is that the social contract operated by a hunter-gatherer community is enforced with a rod of iron. No individual occupies the role of a policeman, but the relatively small size of a hunter-gatherer band makes it possible for *public opinion* to fulfil the same function. When Adam asks himself whether he should offer some of his meat to Eve, he knows very well that he will be relentlessly mocked and ridiculed by the band as a whole should he fail to share in the customary fashion. Full-scale ostracism would follow if he nevertheless persisted in behaving unfairly.

Reports that modern hunter-gatherer communities share on a quasi-utilitarian basis are consistent with the view that public opinion serves as a substitute for an external enforcement agency in such societies. But it is hard to share the enthusiasm expressed by some anthropologists for the oppressive social mechanisms by which discipline is maintained. Envy is endemic. For example, among the !Kung of the Kalahari desert, nobody cares to keep a particularly fine tool for too long. It is passed along to someone else as a gift lest the owner be thought to be getting above him- or herself. But such gifts do not come without strings. In due course, a fair return will be expected. Such close attention to the accountancy of envy in such a social contract makes progress almost impossible. According to Hayek's (1960: 153) definition, the citizens of such a society are free because they are subject to no single individual's will, but it would be a bad mistake for libertarians to idolize such societies. They would do better as a role model for the socialist utopia that Marx envisaged would emerge after the apparatus of the state had withered away.

I therefore diverge from evolutionary psychologists such as Erdal and Whiten who believe that the social contracts of prehistoric hunter-gatherers are preserved in fossilized form by the foraging bands of today. I don't doubt that prehistoric bands were equally free of bosses, but I think it unlikely that they operated a form of social contract that seems to me at least as sophisticated as the authoritarian alternatives operated by ancient tillers of the soil. This claim is of considerable importance to my speculations about the circumstances under which fairness norms evolved, and so it will be necessary for me to defend it at some length.

Farming versus foraging

Cohen (1977) attributes the origins of agriculture to a food crisis in prehistory that arose when human hunter-gatherer bands had expanded until the available habitat was no longer able to support their economies. The response to this over-population problem was twofold. I am particularly interested in the adaptations that allowed foraging to continue in marginal habitats, but anthropologists naturally concentrate on what proved to be the mainstream cultural adaptation — the emergence of agriculture and herding as new modes of production.

The organization necessary both to exploit the increasing returns to scale available in these new modes of production and to prevent the surplus from being appropriated by outsiders made it necessary to abandon the anarchic structure of prehistoric foraging bands. Instead authority was vested in leaders. This readoption of the hierarchical organization typical of ape societies did not require a new set of biological adaptations. We did not lose our capacity to submit to leadership when we acquired the new program that permitted our proto-human ancestors the flexibility necessary to sustain the anarchic lifestyle of hunter-gatherers with a whole world into which to expand. Even in modern foraging societies, our natural urge to dominate our fellows continues to operate in an uneasy relationship with our natural urge to be fair. Otherwise social mechanisms that inhibit dominance behaviour would not be necessary. Anthropologists attribute the social retooling necessary for the transition back to the type of hierarchical social contract needed to maintain a communal farming society to *cultural* evolution. The time available seems too short for a further *biological* adaptation to have been responsible.

It is frequently argued that the human species paid a heavy price for the opportunity to become farmers. When social evolution erected an authoritarian superstructure on a biological foundation that had evolved to permit our ancestors to live a free-wheeling leaderless existence, a war began between part of our biological nature and our social conditioning. Social commentators such as Maryanski and Turner (1992) argue that we are still fighting this war. I express their characterization of a modern industrial society as a *social cage* in the language of game theory by saying that the social conditioning that habituates us to using leadership as an equilibrium selection device conflicts with our natural instinct to employ fairness for this purpose. My guess is that we succeed in tolerating leaders by inventing the social fiction that they are responsible as individuals for the capabilities of the groups they co-ordinate. The worthiness that would be attributed to the group if it were a person is then conferred on its leader. The leader's claim to more than their fair share is thereby rationalized away. But maintaining such a charade is endemically stressful.

Speculative though it is, such a story about the origins of the farming communities from which our own industrial societies are descended seems relatively uncontroversial, but the same is not true of my belief that the social contracts of proto-human foraging bands were too unlike the complex social contracts of modern foragers to allow the analogy to be useful.

Recall Cohen's (1977) suggestion that a prehistoric food crisis caused by over-population spelled the end of foraging as the normal productive mode among humans. But it did not wipe hunter-gatherers out altogether from those old-world territories where the problem arose. Foragers continued to survive in marginal habitats on the fringes of deserts or the polar ice-cap, where growing crops or herding animals is not feasible. Indeed, the fact that such habitats were colonized is one piece of evidence that favours the over-population theory.

To survive in such marginal habitats without the possibility of emigration, foragers had to develop a culture that was no less at variance with their natural instincts than those who took up the farming option. Three cultural adaptations universally observed in modern hunter-gatherer societies seem especially significant. The first is their use of 'natural' methods of birth control — such as the delayed weaning of children. This adaptation goes some way towards solving their population problem. The harshness of their environment when times are bad probably does the rest. The second cultural adaptation was the development of extremely effective social mechanisms that prevent the emergence of leaders or entrepreneurs — except temporarily in emergencies.

Why should mechanisms that inhibit leadership confer an evolutionary advantage? The reason is presumably that innovators are poison for foraging bands occupying marginal habitats. The survival of the memes that regulate the life of a hunter-gatherer society depends on the equilibrium on which its members co-ordinate in a bad year, when food is scarce. If the crisis is sufficiently severe, some members of the band die, and the rest seek refuge with neighbouring bands. But such lean years are infrequent (Sahlins 1974). In the fat years that intervene, memories of the privations of the last lean year will fade. The band will then be at risk of being seduced by a charismatic entrepreneur into co-ordinating on a new equilibrium that does better at exploiting the surpluses available in fat years. Disaster will then ensue if this new social contract is being operated when a lean year comes along.

In brief, the memes that inhibit the appearance of leaders, who are likely to tamper with a traditional social contract tailored to the conditions that prevail in lean years, serve as a kind of collective unconscious that preserves a folk memory of disasters narrowly avoided in the past. The stubborn conservatism of supposedly stupid peasants occupied in subsistence farming doubtless has a similar explanation.

The third cultural adaptation has already been mentioned. Public opinion can serve as a substitute for an external enforcement agency in small close-knit communities. Harsanyi's (1977) analysis of the device of the original position under such circumstances then provides us with an explanation of how the use of the fairness algorithm that I believe is written in our genes can result in quasi-utilitarian sharing of the type reported to be universal among modern hunter-gatherers. (I am suppressing the sceptical streak that reminds me of nineteenth-century reports that the infirm and elderly were abandoned by various nomadic tribes of North America when times became hard. Such behaviour is consistent with my theory, because standards of interpersonal comparison are likely to develop that result in the powerless being deemed unworthy when food is shared. However, such behaviour is not compatible with anthropological reports that food is shared strictly according to need.)

Anarchy in prehistory?

Figure 3 illustrates the speculations about the evolutionary history of modern social contracts offered above. Its significant feature is that the sophisticated social contracts of contemporary hunter-gatherer societies appear on a twig on the branch of the tree that leads to societies like our own. To understand the origins of our instinct for justice, we therefore need to go back to the common ancestor of both types of social contract. But this does not imply that nothing is to be learned from the social contracts of modern foraging bands. They show that human biological hardwiring allows us to operate social contracts without ape-like dominance hierarchies. Since modern hunter-gatherer societies manage without bosses, humans do not need bosses to survive as social animals.

Knauft's (1991) U-shaped curve must therefore be correct insofar as it embodies the claim that the imperative for authoritarianism was somehow cleansed from the genes of our pre-human ancestors, only to be revived in relatively recent history as a *cultural* adaptation to the need to domesticate plants and animals in response to population pressures. To understand the circumstances under which human fairness algorithms evolved, we therefore need to look back to a time after the biological imperative to organize in terms of dominance structures had weakened to an extent that made it possible for more flexible social systems to evolve, but before population pressures had led those societies that continued to forage to develop socially sophisticated methods of controlling both their population as a whole and their selfishly inclined entrepreneurs.

Such prehistoric foraging bands must have differed from their modern descendants in several important respects. In particular, barriers to emigration were absent when the whole world was available for colonization. Under such circumstances, their social organization must have been anarchic to an extent that would make modern hunter-gatherer societies look positively paternalistic. How could it have been otherwise when a dissident group always had the

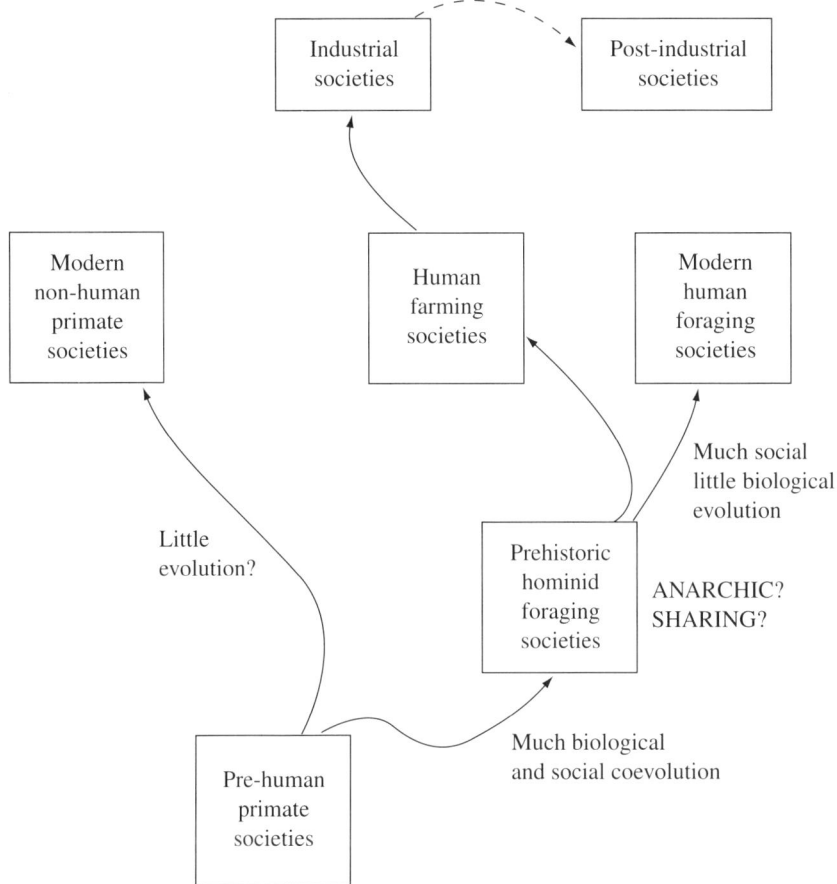

Figure 3. In what environment did fairness evolve?

Lockean option of breaking away at relatively low cost to set up shop in pastures new? Neither public opinion nor personal authority can act as Big Brother when punishment can always be evaded simply by walking off.

With no external enforcement agency available to enforce food-sharing agreements, the Adams and Eves of prehistoric times presumably must have shared food in much the same manner as vampire bats share blood. Vampire bats starve if deprived of blood for more than 60 hours. Wilkinson (1984: 182) observes that close kinship is not necessary for a bat to regurgitate blood for a regular roost-mate. Not only are bats able to recognize each other; the evidence shows that they are more likely to help out a neighbour who has helped them out in the past. Just as each bat in a reciprocating pair has to act as its own policeman in disciplining any tendency by its partner to cheat, so each Adam and Eve who learned to co-operate would have found it pointless to appeal to the rest of the band about any bad behaviour by their partners.

In short, co-operation must originally have been based on *pairwise* interactions, with each pair responsible for policing its own affairs. One can follow Axelrod (1984) in using TIT-FOR-TAT as a representative of the type of reciprocating strategy necessary to support co-operation in such circumstances. However, it is important to bear in mind that it would be a mistake to do the same when discussing societies whose structure cannot be seen as a collection of overlapping *two-person* subsocieties.

Kinship in small groups

Game theorists never tire of warning against proceeding as though the camaraderie that enlivens small groups who work or play together extends to the world in general. The daily life of such small groups awakens the memories buried in our genes of how to interact with others in the small hunter-gatherer communities of prehistory. The brotherhood of man then manifests itself because we are literally programmed to treat each other like brothers and sisters in such circumstances. But Dunbar (1992) has plausibly argued that the size of a group within which it is possible for people to treat each other like family is limited by the capacity of the human neocortex to sustain a social model in which each person in the group and their relationships with others are modelled individually.

This theory about the social dynamics of small groups would seem to be undermined by Erdal and Whiten's (1996) claim that kinship theories fail to explain how food is shared among modern hunter-gatherers. It is doubtless true that a player's share cannot be calculated from his or her family relationships using some simple formula. But it is hard to see how reports that food is shared according to *need* can be explained without assuming that the players actively sympathize with each other's plight, as predicted by Hamilton's rule.

An example will be useful in illustrating the mechanism envisaged. The same example will then be used to compare the food-sharing norms of modern hunter-gatherers with the different food-sharing norms that my theory attributes to the more anarchic hunter-gatherers of prehistory.

Modern hunter-gatherers

Adam and Eve's individual utility functions are normalized so that $u_A(0) = 0$ and $u_A(1) = u_E(1) = 1$. Their shape is determined by the following considerations. Adam and Eve are assumed to be indifferent between obtaining all of the kill brought home by a hunter and receiving some smaller share σ of the kill. In Adam's case, $\sigma = \frac{2}{3}$. In Eve's case, $\sigma = \frac{5}{6}$. Finally, both Adam and Eve are assumed to be risk neutral about shares x that lie between 0 and σ. (The graphs of their individual utility functions therefore consist of two line segments, one joining $(0, 0)$ to $(\sigma, 1)$ and another joining, $(\sigma, 1)$ to $(1,1)$. A better definition of need in this context would replace the first line segment by a line segment joining $(0, 0)$ to $(\sigma, 0)$. A player would then reject any share $x < \sigma$ in favour of any lottery with prizes $x = 0$ and $x = \sigma$. The shares in Table 1 would then be replaced by the probabilities that a player's need is met. However, such a model would have to be complicated by providing players with some form of compensation for settling for a lower probability of winning σ.)

Table 1. Families sharing in modern foraging bands.

Adam	Eve	Adam's share	Eve's share	w_E	w_A	w_E^*	w_A^*
stranger	stranger	½	½	4	5	4	5
cousin	cousin	⁶⁄₁₁	⁵⁄₁₁	3	4	4	5
brother	sister	⅔	⅓	1	2	4	5
son	mother	⅔	⅓	3	10	4	5
father	daughter	⅙	⅚	4	3	4	5

Need is a complex concept, but only one aspect of what it means to say that a person is in need will be relevant here — the extent to which he or she is in want. Adam and Eve's need to eat will be measured by the smallness of the largest share σ for which they are willing to take a risk of ending up with nothing. Their desperation is strengthened by the assumption that they are risk neutral over smaller shares. Since $\frac{2}{3} + \frac{5}{6} > 1$, Adam and Eve's needs cannot be satisfied simultaneously, even though we shall assume that they are the only members of the band who are thought worthy of a share of this particular kill. To what extent will the norm they operate recognize that Adam's need is greater than Eve's?

If I am right about public opinion in a modern foraging band acting as an external enforcement agency, then Adam and Eve's prowess at hunting or gathering will be irrelevant to how the surplus is divided. The social contract will therefore make the lucky hunter's share contingent only on whether that hunter is Adam or Eve. An able hunter may protest and seek to monopolise the kill but will be only one against the combined might of the whole band. The Marxian principle that each should contribute according to his or her ability is therefore realized.

Distribution is not as easy to deal with as production. However, I will proceed as though the band operates a utilitarian norm in which the standard of interpersonal comparison has adjusted until pairs of players divide the surplus according to the Nash bargaining solution. The band as a whole is assumed to enforce its social contract by confiscating the kill should squabbling replace a dignified application of the relevant fairness norm. The relevant *status quo* for the Nash bargaining solution is therefore $(0, 0)$. It is then easy to show that the kill will be split fifty:fifty between Adam and Eve — whatever values of σ we write into their individual utility functions.

This split seems to be egalitarian, but it appears less so when translated into utility terms. The slope of the relevant part of the Pareto-frontier to their bargaining set X is $-\frac{4}{5}$. To implement the outcome in which Adam's share is $x = \frac{1}{2}$ using a weighted utilitarian solution, we can therefore take $w_E = 4$ and $w_A = 5$. Eve's lesser need is then reflected in her individual utils being counted as worth only $\frac{4}{5}$ of Adam's. But since the moral content of a fairness norm is eroded away as w_A and w_E adjust until $h = n$ in Figure 1(b), this difference in their perceived worthiness does not result in a move in Adam's direction away from the fifty:fifty split.

Now alter the story by making Adam and Eve relatives. Their individual utility functions then have to be replaced by personal utility functions that incorporate Hamilton's inclusive fitness criterion. If Adam's degree of relationship to Eve is r, his individual utility $u_A(x)$ for a split of the kill in which he gets x and Eve gets $1 - x$ must be replaced by his true personal utility:

$$v_A(x) = u_A(x) + r u_E(1 - x).$$

Similarly, if Eve's degree of relationship to Adam is s, then $u_E(x)$ must be replaced by $v_E(x) = u_E(x) + s u_A(1 - x)$.

The inbreeding that is inevitable in small isolated groups will be ignored in the first instance. When Adam and Eve are siblings, gene-counting arguments imply that $r = s = \frac{1}{2}$. When they are first cousins, $r = s = \frac{1}{8}$. The possibility that $r \neq s$ is included to take account of parent–child relationships. If a son has survived until puberty and his mother is no longer nubile, then the degree of their relationship has to be altered to take account of their different chances of reproducing their genes. In this example, I consider the extreme cases when $r = \frac{1}{2}$ and $s = 0$, and $r = 0$ and $s = \frac{1}{2}$.

Table 1 shows how the kill is divided when family relationships are taken into account. The constants w_E and w_A are weights whose use ensures that the agreed split maximizes the weighted utilitarian solution, provided that Adam and Eve's personal utilities are properly evaluated to show their sympathy with each other. The constants w_E^* and w_A^* perform the same function in the case when Adam and Eve's personal utilities are mistakenly replaced by their individual utilities.

Except when Adam is Eve's father, the closer the relationship between Adam and Eve, the more his greater need is recognized. When he interacts with his mother or his sister, his needs are met in full. Such recognition of Adam's need is also evident in the standards of interpersonal comparison that operate in the different cases. For example, when Adam and Eve are siblings, one of

Adam's utils is deemed to be worth only half of one of Eve's. When they are cousins, his utils are worth three-quarters of hers. In the exceptional case when Adam is Eve's father, his unreciprocated concern for her welfare results in her needs taking total precedence over his. Her utils are then deemed to be worth three-quarters of his.

Although no simple formula connects who gets what with how they are related, kinship clearly provides a good explanation of why the needy receive special treatment in modern hunter-gatherer societies. The phenomenon is strengthened if we take account of the fact that inbreeding will increase the degrees of relationship. A simple model that assigns probability ρ to the event that any married couple share a particular gene, attributes a degree of relationship $r = \frac{1}{2} + \frac{1}{2}\rho$ to siblings, and $r = \frac{1}{2} + \frac{7}{8}\rho$ to cousins. If $\rho = \frac{1}{7}$, then Adam's needs are met in full, although Adam and Eve may have no obvious family connection at all.

An observer will then see all members of the same generation sharing food as though only need matters. But one cannot deduce that kinship is irrelevant to the way food is shared. On the contrary, the needy are cared for *because* they are kin.

Prehistoric hunter-gatherers

I have argued against assuming that modern foraging bands will serve as a model of the prehistoric foraging bands of our ancestors. If I am right, the social contracts of prehistoric hunter-gatherers were enforced neither by a powerful leader nor by the whole group acting in concert. The parties to a sharing agreement therefore had to police the deal themselves. A form of social organization in which each citizen produces according to his or her ability and consumes according to his or her need would have been beyond their comprehension.

In the simple example we have been studying, the difference between the two forms of social contract emerges in the location of the state of nature. In a prehistoric foraging band, the band as a whole would not have disciplined Adam and Eve by confiscating their product if they fought over its division instead of operating the conventional fairness norm. The analysis therefore needs to be modified so that the *status quo* used when applying the Nash bargaining solution becomes some analogue of Buchanan's (1975) *natural equilibrium*. (My theory predicts that prehistoric foragers used the egalitarian bargaining solution rather than the weighted utilitarian solution, but the result will be the same in both cases after cultural evolution has finished operating on the worthiness coefficients.)

I assume that the probability that the kill is left in Adam's hands after a failure to agree on an insurance contract is $p = \frac{1}{5}$. In the cleanest case, there is no

Ken Binmore

fighting and a failure to agree simply leaves the carcase in the hands of the player who made the kill. With no fighting, we can identify p with the probability that Adam is the successful hunter, and so p serves as a measure of ability.

The parameter choices in the model imply that we are to study the case in which Eve is more able and Adam is more needy. Table 2 compares the shares each now receives with the case of a modern hunter-gatherer society. Notice that the standards for making interpersonal comparisons have not changed, but the new power structure in their game of life dramatically alters Adam and Eve's share of the surplus. Only Eve's needs are satisfied when Adam and Eve are no more related than cousins. When Eve is Adam's mother, she still reserves more than half the surplus for herself. Even if a high level of inbreeding with $\rho = \frac{1}{3}$ is postulated, Adam still gets less than $\frac{4}{7}$ of the surplus when Eve is as closely related as a sister.

Table 2. Families sharing in prehistoric foraging bands.

Adam	Eve	Adam's share	Eve's share	w_E	w_A	w_E^*	w_A^*
stranger	stranger	$\frac{1}{6}$	$\frac{5}{6}$	4	5	4	5
cousin	cousin	$\frac{1}{6}$	$\frac{5}{6}$	3	4	4	5
brother	sister	$\frac{1}{3}$	$\frac{2}{3}$	1	2	4	5
son	mother	$\frac{7}{15}$	$\frac{8}{15}$	3	10	4	5
father	daughter	$\frac{1}{6}$	$\frac{5}{6}$	4	3	4	5

CONCLUSION

This paper has first offered a summary account of a theory of fairness developed in a much more leisurely style in Binmore (1994, 1998) and then discussed the relevance of the available anthropological data on modern hunter-gatherers to the evolutionary assumptions of the theory. To sustain the theory, it turns out to be necessary to believe that modern hunter-gatherers operate considerably more sophisticated social contracts than did our primitive ancestors.

This paper is a digest of ideas that are explained at greater length in my book *Game Theory and the Social Contract, Vol. II: Just Playing* (1998).

Note. The support of the ESRC Centre for Economic Learning and Social Evolution and of the Leverhulme Trust is gratefully acknowledged.

REFERENCES

ADAMS, J. 1963: Towards an understanding of inequity. *Journal of Abnormal and Social Psychology* 67, 422–36.

ADAMS, J. 1965: Inequity in social exchange. In Berkowitz, L. (ed.), *Advances in Experimental Social Science* (New York, Academic Press), Vol. II.

ARISTOTLE. 1985: *Nicomachean Ethics*, trans. T. Irwin (Indianapolis, Hackett).

AUMANN, R. & MASCHLER, M. 1995: *Repeated Games with Incomplete Information* (Cambridge, MA, MIT Press).

AXELROD, R. 1984: *The Evolution of Cooperation* (New York, Basic Books).

BAILEY, R. 1991: The behavioral ecology of Efe pygmy men in the Atari Forest, Zaire (Technical Report Anthropological Paper 86, University of Michigan Museum of Anthropology).

BINMORE, K. 1994: *Game Theory and the Social Contract, Vol. I: Playing Fair* (Cambridge, MA, MIT Press).

BINMORE, K. 1998: *Game Theory and the Social Contract, Vol. II: Just Playing* (Cambridge, MA, MIT Press).

BUCHANAN, J. 1975: *The Limits of Liberty* (Chicago, University of Chicago Press).

BYRNE, R. & WHITEN, A. 1988: *Machiavellian Intelligence* (Oxford, Clarendon Press).

COHEN, M. 1977: *The Food Crisis in Prehistory: Overpopulation and the Origins of Agriculture* (New Haven, Yale University Press).

DAMAS, D. 1972: Copper Eskimo. In Bicchieri, M. (ed.), *Hunters and Gatherers Today* (New York, Holt, Rinehart & Winston).

DUNBAR, R. 1992: Neocortex size as a constraint on group size in primates. *Journal of Human Evolution* 20, 469–93.

ERDAL, D. & WHITEN, A. 1996: Egalitarianism and Machiavellian intelligence in human evolution. In Mellars, P. & Gibson, K. (eds.), *Modelling the Early Human Mind* (Oxford, Oxbow Books).

EVANS-PRITCHARD, E. 1940: *The Nuer* (Oxford, Clarendon Press).

FURBY, L. 1986: Psychology and justice. In Cohen, R. (ed.), *Justice: Views from the Social Sciences* (Cambridge, MA, Harvard University Press).

GARDNER, P. 1972: The Paliyans. In Bicchieri, M. (ed.), *Hunters and Gatherers Today* (New York, Holt, Rinehart & Winston).

HAMILTON, W. 1963: The evolution of altruistic behavior. *American Naturalist* 97, 354–6.

HAMILTON, W. 1964: The genetic evolution of social behavior, parts I and II. *Journal of Theoretical Biology* 7, 1–52.

HARSANYI, J. 1977: *Rational Behavior and Bargaining Equilibrium in Games and Social Situations* (Cambridge, Cambridge University Press).

HAWKES, K., O'CONNELL, J. & BLURTON-JONES, N. 1993: Hunting income patterns among the Hadza. In Whiten, A. & Widdowson, E. (eds.), *Foraging Strategies and Natural Diet of Monkeys, Apes and Humans* (Oxford, Clarendon Press).

HAYEK, F. 1960: *The Constitution of Liberty* (Chicago, University of Chicago Press).

HELM, J. 1972: The Dogrib Indians. In Bicchieri, M. (ed.), *Hunters and Gatherers Today* (New York, Holt, Rinehart & Winston).

HOMANS, G. 1961: *Social Behavior: Its Elementary Forms* (New York, Harcourt, Brace & World).

HUME, D. 1978 (1739): *A Treatise of Human Nature* (*Second Edition*), edited by L.A. Selby-Bigge, revised by P. Nidditch (Oxford, Clarendon Press).

ISAAC, G. 1978: The food-sharing behavior of protohuman hominids. *Scientific American* 238, 90–108.

KAPLAN, H. & HILL, K. 1985: Food sharing among Ache foragers: tests of explanatory hypotheses. *Current Anthropology* 26, 223–45.

KNAUFT, B. 1991: Violence and sociality in human evolution. *Current Anthropology* 32, 223–45.

LEE, R. 1979: *The !Kung San: Men, Women and Work in a Foraging Society* (Cambridge, Cambridge University Press).

MARYANSKI, A. & TURNER, J. 1992: *The Social Cage: Human Nature and the Evolution of Society* (Stanford, Stanford University Press).

MEGARRY, T. 1995: *Society in Prehistory: the Origins of Human Culture* (London, Macmillan).

MEGGITT, M. 1962: *Desert People: a Study of the Walbiri Aborigines of Central Australia* (Chicago, University of Chicago Press).

MELLERS, B. 1982: Equity judgment: a revision of Aristotelian views. *Journal of Experimental Biology* 111, 242–70.

MELLERS, B. & BARON, J. 1993: *Psychological Perspectives on Justice: Theory and Applications* (Cambridge, Cambridge University Press).

RAWLS, J. 1972: *A Theory of Justice* (Oxford, Oxford University Press).

RICHES, D. 1982: Hunting, herding and potlatching: toward a sociological account of prestige. *Man* 19, 234–51.

ROGERS, E. 1972: The Mistassini Cree. In Bicchieri, M. (ed.), *Hunters and Gatherers Today* (New York, Holt, Rinehart & Winston).

SAHLINS, M. 1974: *Stone Age Economics* (London, Tavistock).

SELTEN, R. 1978: The chain-store paradox. *Theory and Decision* 9, 127–59.

TANAKA, J. 1980: *The San Hunter-Gatherers of the Kalahari Desert: a Study of Ecological Anthropology* (Tokyo, Tokyo University Press).

TRIVERS, R. 1971: The evolution of reciprocal altruism. *Quarterly Review of Biology* 46, 35–56.

TURNBULL, C. 1965: *Wayward Servants* (London, Eyre & Spottiswoode).

WALSTER, E., WALSTER, G. & BERSCHEID, E. 1978: *Equity: Theory and Research* (London, Allyn & Bacon).

WILKINSON, G. 1984: Reciprocal food-sharing in the vampire bat. *Nature* 308, 181–4.

WILSON, J. 1993: *The Moral Sense* (New York, Free Press).

Evolutionary Perspectives on the Origins of Human Social Institutions

ROBERT A. FOLEY

Social institution — 'a fairly permanent cluster of social usages. It is a reasonably enduring, complex integrated pattern of behaviour by which social control is exerted and through which basic social desires or needs can be met.'

Winick, *Dictionary of Anthropology,* 1960

INTRODUCTION

IF IT CAN BE BROADLY ACCEPTED that human social institutions have an evolutionary origin, then there are two possible models of how they may have evolved. One such model would see the origin of any institutions as resting in the nature of the human mind and cultural capacity, and therefore having a history that would run parallel to the evolution of the human mind in general. This would imply, on the whole, a relatively long evolutionary history, certainly as long as that of the human species. The other potential model would place the emphasis not on the innate capacities of the human species, but on the specific context in which humans find themselves. Social institutions would arise and disappear in response to specific conditions. This model is much more flexible and context-specific, and would posit a rather shorter and more variable history for human social institutions.

Clearly, which of these two extreme models is adopted will depend to a large extent on the nature and definition of the social institutions concerned. However, they also broadly represent two views of the nature of human social evolution: one such view, associated with the emerging field of evolutionary psychology, tends to emphasize the underlying and presumably genetic basis of social behaviour, and lays stress on its universality and deep evolutionary origins (e.g. Barkow *et al.* 1992). The other is far more aligned with a view that sees human behaviour as highly variable, shaped by socioecological context, and liable to produce more transient social institutions (e.g. Borgerhoff Mulder 1996; Hinde 1987).

In the end, of course, both views must be incorporated into any synthesis of the evolution of human social institutions. There must be both a set of human

Proceedings of the British Academy, **110,** 171–95, © The British Academy 2001.

behavioural propensities, shaped by selection, and a pattern of historically specific developments, shaped by demographic and ecological circumstances. Both elements are an essential part of any evolutionary reconstruction.

The aim of this chapter will be to examine some particular social institutions in this light. In particular, I shall develop a model in which the community, with a probable set of internal social relationships, is the fundamental unit of human social organization, and which states that it is the potential for communities to vary in relation to demography and ecology that sets the parameters for the evolution of other social institutions. The chronological context is the long-term evolution of the human lineage, and it will be argued that different social institutions have very different histories — some stretching back to the first members of the genus *Homo*, some to more recent evolutionary changes, and some to the demographic expansion of human populations in the past 10,000 years.

HUMAN SOCIAL INSTITUTIONS

Perhaps the first question that should be addressed is: what are human social institutions? As the quotation at the start of this chapter shows, this is not one to which there can be a concrete answer. Because evolutionary analyses require a relatively simple and broad-brush approach, largely because of the crude nature of the information available (for example, archaeological), a simple definition will be provided here. Humans, along with the anthropoid primates, are innately and compulsively social. They live in groups in which individuals develop stable relationships and have the behavioural, cognitive, and emotional capacity to tolerate the close proximity of other individuals. Social institutions may be considered to be those relatively persistent structures which emerge from the way this capacity is put into practice in particular social contexts. They thus 'reside' in both the mental structures of individuals and their actual interactions with others, and have the effect of co-ordinating behaviour, maintaining order, and enhancing the well-being of either the whole or parts of any group. Any investigation of the origins of social institutions must therefore take into account both the potential inherent in the species and the shifting context.

Table 1 outlines what I consider to be some primary human social institutions. The list is to some extent arbitrary, and no doubt others could be included. These have been selected to demonstrate the principle that social institutions are hierarchically organized in relation to, on the one hand, substructure within communities, and, on the other, relationships between communities. Thus in terms of the community itself, institutions would include those that might promote egalitarianism, such as councils, and those that

Table 1. Human social institutions.

Family level	Community level	Inter-community level
Marriage patterns	Social mechanisms for	Political organization
Kinship systems	promoting egalitarianism	Warfare
Descent groups	and consensus	Trade/exchange
Resource distribution	Hierarchies and 'chiefs'	Religion?
(sharing)	Lineages and descent groups	
	Social norms (taboos, etc.)	
	Law, coercion, and punishment	
	Resource distribution	
	Marriage rules (incest avoidance)	
	Religion/cosmology	

Note: In the text it is proposed that social institutions are primarily located in the community, and these are extended both to sub-groups within the community (e.g. families) and to relationships between groups. The list here is not exhaustive, but highlights the primary social institutions discussed in this chapter.

enhance leadership, such as hierarchies and chieftainships; they would also comprise the lineages and descent groups around which they will normally be organized. In a slightly different form they would also include the mechanisms by which the norms of social behaviour are established and maintained, such as taboos, rules of marriage, laws, methods of coercion and punishment, and mechanisms for resource distribution (see Table 2). Underpinning all of these will be systems of cosmology, religious belief, and moral order.

These community-level institutions will be replicated in the sub-structures that exist within the community, for example among families and descent groups, and variations will emerge at that level — for example, different families may operate different rules of behaviour for the distribution of resources or operate different systems of coercion and punishment from those found more broadly across the community.

Social institutions can also be said to exist beyond the community. Except in exceptional circumstance of complete isolation, all communities must exist in a relationship with those around them. These relationships will involve both competition and co-operation, and will be reflected in systems of political organization, patterns of trade and exchange, and the potential for or reality of warfare. In effect, the aim of this chapter is to provide some level of explanation for the emergence and diversity of these institutions. Before doing so I shall briefly discuss the nature of the human community, and the proposition that from an evolutionary perspective this constitutes the basic unit and starting point for considering other elements of social behaviour.

Table 2. Main social parameters that vary within a community and give rise to differences in social institutions.

Community size	Group structure	Family organization	Kinship	Marriage	Mobility
Variation in size from small isolated bands and autonomous families to cities $10-10^6$	Hierarchical Egalitarian Sub-structured	Nuclear families Extended families Patterns of co-residence Matrilocal, patrilocal, etc.	Patrilineal Matrilineal Bilateral Depth of descent groups	Monogamous Polygynous Polyandrous Unstable/ promiscuous	Nomadic Fission-fusion Tethered transhumance Permanent settlements

THE CONTEXT FOR THE EVOLUTION OF HUMAN SOCIAL INSTITUTIONS

The human social community

Comparative primatology shows that all anthropoid primates live in social groups, and thus these groups and their inherent sociality constitute the ancestral state for any considerations of humans. The structure of primate groups is diverse, but the comparative method would suggest focusing on the closest living relatives of humans, the chimpanzee and bonobo. Both these species have been described as living in multi-male, multi-female groups that persist over time; these have been referred to as *communities* (Goodall 1986). Using the finite social space model, Foley and Lee (1989) placed humans in the same multi-male, multi-female community structure, and Rodspeth *et al.* (1991) extended this model to show that the diversity of human social organization can be subsumed within this community-based approach. Alternatives such as bands, tribes, etc. are either too specific, or else have specific implications about structure, which are unwarranted when applied to the totality of human social organization.

I would suggest that the community, defined in this way, is the fundamental unit of human social organization. Community structure reflects the ties of kinship, the prolonged nature of social interactions between adult males and adult females, the units within which offspring are born and raised, and also the units in which nuclear families must be embedded. In some cases this will be synonymous with the band, where there may be no significant larger grouping; in other cases the band may be a sub-unit within a larger grouping, such as the tribe, which may constitute a community itself. For hunter-gatherers and pastoralists these may be residentially flexible, but with the establishment of sedentary ways of life, particularly with agriculture, then the community may be a village or even a town. However, among large urban aggregations the communities would be sub-units within them.

Socioecology of human communities

If communities are the fundamental unit of human social organization, and the locus for social institutions, then it follows that these social institutions will arise from the particular characteristics of the community, the way it operates and functions in particular contexts. The basic thesis developed in the second half of this chapter is that the major human social institutions have emerged as adaptive solutions to the particular contexts in which human communities found themselves. In order to understand how particular institutions arose, it is necessary to develop a theoretical framework which will allow us to explore how community structure will vary, and thus how social institutions are a response to this variation.

The framework used here is essentially drawn from evolutionary theory and, in particular, behavioural ecology (also known as socioecology). This branch of evolution and ecology sets out to explain how patterns of behaviour, including social behaviour, relate to resources and resource distribution, and has been used extensively in zoology to account for variation in behaviour, within and between species (Krebs & Davies 1997; Lee 1999; Standen & Foley 1989), and has also been used in anthropology and sociology to account for patterns of behaviour and culture (Betzig *et al.* 1988; Layton *et al.* 1991; Runciman 1998; Smith & Winterhalder 1992). Only a few key points will be made here. The central point of socioecology is that the way resources are distributed and the costs of exploiting them are the key determinants. The resources themselves are not inherently valuable or costly, except in relation to the strategies involved and the competitive context. For example, a rabbit may be an extremely costly resource for a human to chase and catch with bare hands, but with a trap it can become a highly valued resource. Equally, as the density of resources declines through either environmental change or over-exploitation, as has happened many times in prehistory, the resource may go from being worth exploiting to irrelevant. This is the principle underlying optimality theory, which is the basis for much socioecology. The reason for stressing this point here is that it emphasizes the dynamic nature of behavioural ecology, and hence its value in studying patterns of evolutionary change in behaviour.

Turning to the more substantive question of the socioecology of human communities, the starting point should be the small-scale communities of hunter-gatherer bands and their analogues in pre-modern populations, from which more complex communities emerge. If we look at variation in this (see Hayden 1981 and Kelly 1995 for reviews, and Foley & Lahr in press for a discussion of ecological variation in relation to demography and genetics), a number of patterns emerge. For hunter-gatherers at least, but other communities as well, the major dimensions of variation are the size of the community and the area over which it is distributed. Communities can be very small, a few tens of people, or much larger, more than one thousand. They can also vary in the extent to which they are densely packed in small areas, such as is the case for Andaman Islanders, or range over thousands of square kilometres, as is the case for most Inuit populations. Table 3 summarizes the pattern of variation. There is a strong resource and environmental basis to this pattern. Where resources are localized, patchy, predictable, and of high quality, then communities have the potential to become larger and/or to be densely packed in small areas. The ethnographic and archaeological evidence suggests that this occurs when communities are dependent upon aquatic resources, exploiting large herds of game with predictable distributions, and with the onset of food

Table 3. Socioecological variation in human community size and structure, indicating range of variation and ecological factors that will increase or decrease the size and distributions of communities (from Foley & Lahr in press).

Variation and ecological pattern	Community size	Community range area	Population density	Mobility
Ultimate effect of increase	Permanent community fission Formation of cultural, linguistic and genetic boundaries	High levels of community fissioning	Overcrowding Fission of communities Dispersals and geographical expansion Inter-group conflict	Low reproductive rates?
Conditions leading to upper limits	Larger communities Dependence upon aquatic resources Increased sedentism Richer resources Territoriality and inter-group conflict Population growth Agriculture	Larger ranges Dependence upon hunted animals Scarce and sparse resources High latitudes	High population density Agriculture Aquatic resource dependence Rich resource base Mid-latitudes	High mobility Scarce and dispersed resources High levels of hunting Seasonal environments
Upper limits	500–1000 (per km^2)	10,000–25,000 (per km^2)	>50 per km^2	>50
Modal value	150–250 individuals (per km^2)	1,000–2,000 (per km^2)	1 per km^2	10–20
Lower limits	30 (per km^2)	100 (per km^2)	0.1 per km^2	0
Conditions leading to lower limits	Smaller communities Sparse, scarce resources High levels of mobility Large range area Population decline Community fissioning	Smaller range areas Islands Aquatic resource dependence Equatorial latitudes Plant dependence High population density Rich resources Agriculture	Low population density Deserts, arctic, and resource-poor areas Large game hunting Resource depletion	Low mobility Agriculture Aquatic resource dependence High population density and packing
Ultimate effect of decrease	Community extinction Fusion with other communities			Formation of settled communities in villages and towns Resource depletion

production, and this is consistent with patterns found more widely in the animal kingdom. The social consequences of what happens to communities that grow beyond the hunter-gatherer median of between 150 and 250 will be discussed later, as will the implications of smaller range areas and higher population densities.

Integrally related to group size and range area is the matter of mobility. Some hunter-gatherer communities are highly mobile (see Table 3), moving in some cases almost weekly. There is a pattern of variation from this level to completely sedentary groups, such as those in the north-west Pacific region of North America. Again this pattern reflects resource distribution, with sedentism becoming more pronounced with densely packed resources, renewable resource (such as cereals), and aquatic resources. Fission and fusion of groups is an important part of the social behaviour of many communities, and the absence of this pattern of behaviour is one of the major factors leading to changes in social behaviour among humans

The two aspects discussed above relate to how the population as a whole is distributed in time and space, and how this reflects resource factors. Resources will also affect the way the community itself is structured. There is an extensive literature on this, showing the many complexities (e.g. Betzig *et al.* 1988), but only one element will be discussed here — that of residence patterns.

Communities are not random associations of individuals, but strongly structured. The fundamental basis for these communities is kinship; while this refers primarily to genetically based relationships, it has been well established by anthropologists that kinship systems are not simple reflections of genetic relatedness, and nor are they necessarily confined to such relationships. However, perhaps the key element in kinship systems that is relevant here is the lineage or the descent group. What is striking is that, among hunter-gatherers at least, patrilocality is the most prevalent system (Kelly 1995); that is, upon marriage or at maturity, females are more likely to disperse to other communities, and males to remain resident. This is also a pattern found in chimpanzees, and it has been argued by several authors that this is the ancestral human condition and predominates through our evolution (Foley 1987, 1989; Foley & Lee 1989; Wrangham 1987). This structure has a number of important implications for such aspects as inter-group relationships that will be discussed below. In particular, there is a level of hostility between many groups (Wrangham & Peterson 1997), although this paradoxically will occur in the context of other more pacific relationships and the exchange of marriage partners.

The socioecological basis for patrilocality and male kin-bonding is related to reproductive strategies. Wrangham (1980) argued that females, with their high reproductive costs, are constrained by resource distribution, and will spread themselves across a landscape to optimize their access to resources. Males, on the other hand, are constrained in their reproductive success not by access to resources, but by access to females; they will thus distribute them-selves in relation to the females. This pattern will thus result in any form of social organization from monogamy — dispersed females, with males able to access only one female — and polygyny — where the females are in larger groups, and males can defend them — to multi-male, multi-female systems. Male residence, it has been argued for the chimpanzee (Wrangham 1986), occurs where the females are in groups, but these groups are wide ranging, and so males on their own are unable to control access to females; under those cir-cumstances, they can form coalitions with other males (relatives), and thus defend a community of females as a group. This pattern seems to be related to relatively large-bodied, long-lived species, where there are considerable benefits to the males accruing from long-term maintenance of reproductive activity. This pattern is likely to hold strongly for humans as well (Foley 1989; Foley & Lee 1995).

Finally, it should be noted that underlying all patterns of community structure are mating and parenting: the community persists through time because individuals are able to reproduce successfully and to bring up offspring. Table 4 shows the basic reproductive parameters for hunter-gatherer popula-tions. Once again, variation in these parameters is sensitive to resource avail-ability, and it has been extensively argued that the demographic changes associated with aquatic exploitation, sedentism, and food production lead to higher reproductive rates. However, these are probably extremes of a pattern of variation that can be found in all human communities.

To summarize, human communities vary in size and structure in relation to resource distribution and the means used to exploit resources. These shape social structure, but the form these structures take (fissioning communities, inter-group hostility, marriage patterns, etc.) will in turn have an effect on the access of particular groups of individuals or other communities to those resources, thus setting in train a dynamic process. It is this dynamic aspect that will be explored in the next section.

A MODEL FOR THE EVOLUTION OF SOCIAL INSTITUTIONS IN THE CONTEXT OF RECENT HUMAN EVOLUTION

The discussion of human social community structure and variation presented above suggests three conclusions. The first is that, regardless of its size, the community is the context in which more complex social institutions must have their origin and their rationale; social institutions evolved to mediate stable relationships and functions within and between communities. We must look, therefore, at how the community itself has evolved over time to understand the conditions under which particular institutions may have developed. The

Table 4. Variation in reproduction and mortality among hunter-gatherers, and ecological factors influencing variation.

Parameter	Range	Approximate modal value	Factors affecting variation
Age of first reproduction (female)	14–18 years	17	Primarily social factors, although high levels of nutrition will lower the age of sexual maturity.
Age of first reproduction (male)	18–30 years	26	Primarily social factors; young males will be excluded from reproduction where resource acquisition is an important prerequisite.
Fertility rates	3.7–9.7 live births		Nutrition and disease; agriculture may marginally increase birth rates, but sedentism may also lead to higher sterility rates; high mobility.
Infant mortality	?	20%	Infant mortality will increase where disease levels are high, and this may increase with population density and agriculture.
Pre-adult mortality	?	40%	Childhood mortality is probably largely affected by risk — accidents and violence — which may be higher among hunter-gatherers.
Female reproductive variance	3–7	?	Primarily affected by the availability of resources to females, and will increase with social stratification.
Male reproductive variance	9–12	?	Primarily affected by the social factors relating to male — male competition; variance will increase in societies with greater levels of social stratification. Male reproductive success is strongly influenced by longevity, and will increase where there are high differentials in male mortality.

Source: Foley & Lahr in press.

second is that the male kin-bonded lineage must be central to any reconstruction of the pattern of human social evolution; this was the probable ancestral condition, is the most frequent in contemporary and ethnographically observed societies, and its presence has major implications for relationships between groups. As it is the growth of such interactions that is central to the discussion here, it follows that many institutions will have their origins in the outcomes of male kin-bonded lineage systems, including their transformation into other systems. The third conclusion is that as ecological conditions change, whether they are brought about by natural environmental change or shifts in human subsistence strategies, the size and structure of human communities will also change, and it is these changes that will lead to the development of new social institutions.

The ancestral hominin condition and its evolution

The implication of phylogeny is that early hominins lived in multi-male, multi-female communities, in which males remained resident and females dispersed at maturity (Foley 1987, 1989; Ghiglierhi 1987; Wrangham 1987). How would this basic condition change over the course of long-term, Pleistocene, human evolution? I have argued elsewhere (Foley 1987, 1989, 1995; Foley & Lee 1989, 1991, 1995) that the development of meat eating among early representatives of the genus *Homo* would have been a critical shift, leading to a number of social changes. The first of these changes would have been greater spatial ranging, and this is likely to have expanded the level of fissioning and fusioning compared with that seen in chimpanzees; rather than this occurring on an hourly and daily basis, greater foraging distances would have led to the formation of sub-groups that may have been independent for longer periods. This would have been the first step towards the maintenance of 'communities at a distance' — a characteristic of human groups, where membership of a social community depends not simply upon day-to-day contact, but on stored memory and previous experience. In a sense, this represents one of the first social institutions — the persistence of the communities through time regardless of spatial proximity — and it is likely to be one of considerable antiquity.

A second change relates to the nature of meat as a high-quality resource. Meat would have provided more energy for the mother, and hence reduced the costs of encephalization (Aiello & Wheeler 1995; Foley & Lee 1991); greater encephalization would have led to a more prolonged life history strategy, with a greater level of maternal effort. In social terms one of the most probable effects would have been greater affiliation between males and females. This statement is dependent upon the hypothesis that it was primarily the males who were hunting. The basis for this hypothesis is that among both chimpanzees and modern hunter-gatherers males are by and large the exclusive hunters (Lee

& DeVore 1968; Stanford 1999). Although we cannot rule out the possibility that this was not the case in the past, it is the most likely. If males are hunting and if females are gaining access to the meat, then this could be a factor in the closer, less promiscuous (or less openly promiscuous) relationships between males and females in the hominin lineage. This suggestion would gain some support both from primatological observations that there is an 'exchange' of food for sex among hunting chimpanzees (Stanford 1999), and from observations that among some contemporary hunter-gatherers there is a relationship between hunting prowess and access to females (Kaplan *et al.* 2000). In terms of social institutions, what is occurring here is the development of more exclusive relationships between particular males and particular females. While this would, in its formative stages, be a long way from what may be considered to be 'marriage', and still further from exclusive pair-bonding (for there is every reason to see these relationships as polygynous), none the less it would have been an important initial element in the development of institutions governing the relationships between males and females (Lovejoy 1981).

The third shift in relation to meat eating that can be postulated is related to life history theory. Early *Homo* is associated with the development of slower growth rates (but still accelerated relative to modern humans), and presumably associated changes in other life history parameters (Hammer & Foley 1996; Smith 1989, 1992). In particular, maximum longevity may well have been increased. If this were the case, then the opportunity for communities to include several generations at one time would be greater. The presence of multiple generations would have prompted the existence of descent-based lineages, which, it was argued above, form a key element of the social structure of many contemporary human societies (see O'Connell *et al.* 1999 and Foley 1994 for different interpretations of the implications of this development).

One can argue that the ancestral condition on which later social evolution was based contained elements that go back to the common ancestor with *Pan*, namely the basic community, male kin-bonding, and presumably a high level of territoriality, but also a number of novel elements — the ability to maintain communities over a distance, stronger affiliation between males and females, and rudimentary descent groups. These three traits were important to the later emergence of other social institutions.

After the appearance of *H. ergaster*, and its behavioural contrasts with the australopithecines, there is a major problem. Between around 1.5 and 0.4 Myr there is very little observable change, other than an expansion of habitats occupied, perhaps the development of fire, and an increase in brain size (see Klein 1999 for a review of this evidence). While this might be interpreted as social stasis, and may well have been, it may also simply reflect the fact that the archaeological and fossil record for this period is very fragmentary and patchy. It could therefore be that the social institutions inferred to be in place among *H. ergaster* did not give rise to much further change, or that we have not yet been able to identify these changes.

Whichever is the case, there is evidence for more of a change after approximately 250,000 years ago. There is considerable controversy as to the nature and timing of this shift, but the descendent forms of hominin concerned are the Neanderthals and modern humans. In both these species there is evidence for similarly larger brains (within the range of modern humans), modern growth parameters or ones close to those of modern humans (Dean *et al.* 1986), possibly language (McLarnon 1996), similar and complex technology (Mode 3 or prepared core technology) (Foley & Lahr 1997), and a more efficient projectile-based form of hunting and gathering (Stiner *et al.* 1999). A further change, which is probably associated only with modern humans, and may be of considerable importance, is the greater use of aquatic resources. The evidence from Klasies River Mouth, for example, which has some of the earliest representatives of modern humans, suggests that at this stage midden formation was occurring, and this would be evidence for major dependence upon this resource (Deacon 1989; Deacon & Shuurman 1992; Klein & Cruz-Uribe 1996).

It is very hard to put together a coherent story for human social institutions at this stage. At one level, the last common ancestor of Neanderthals and *H. sapiens* (*H. helmei* according to one model, *H. heidelbergensis* according to another) may well have differed in several ways from modern humans, but across this time (i.e. between 250,000 years ago and 100,000 years ago), fully modern cognitive and behavioural abilities must have come into place. When and how this occurred is far from clear. However, early modern human sites in southern Africa 100,000 years ago suggest that there was a greater population density among at least some communities (Deacon & Shuurman 1992). This may be associated with these populations being more sedentary through use of aquatic resources. If the model of male kin-bonded groups is correct at this stage, then one implication is that there may have been a trend towards greater territorial conflict, high rates of community fissioning, and perhaps a tendency towards dispersal of communities and territories. This would certainly fit the current evidence for the dynamic process by which modern humans emerged and colonized the world during the later part of the Pleistocene. The evidence for territoriality and conflict is extremely patchy, but it is suggestive that several of the later Pleistocene hominins (Klasies River Mouth) do show traumas which are consistent with violent deaths or at least injuries (White 1987). However, the causes of these are unknown.

Perhaps the key point to emphasize at this stage is the following. All the evidence from genetics and the fossils indicates that the modern human species evolved at some time between 200,000 and 100,000 years ago. There is, however, very little archaeological evidence which would point to the emergence of any new social institutions at this time (but see Brooks *et al.* 1995 for some evidence relating to bone points). It is only considerably later, after fully modern biological features and abilities were in place, that such evidence begins to appear (Klein 1992, 1995, 2000; Mellars 1989, 1996). How can this best be interpreted? Perhaps the most parsimonious explanation is that while the evolution of modern humans, or indeed perhaps of their common ancestors with the Neanderthals, put into place behavioural and cognitive modernity, this did not lead to any major social change because there was a missing component. That component was not the innate biological characteristics of the human species, but the ecological context which would transform humans from a sparsely dispersed, low-density population to one where larger and larger communities, with considerable complexity of social form, would become increasingly common. The factors underlying this will be considered in the next section.

Phase 1: the early social history of *Homo sapiens*

I have outlined how some elements of human social institutions may have either evolved or existed in some form during the course of the evolution of *Homo* over a period of up to two million years. However, the absence of concrete evidence for any complexity suggests that most social institutions do not occur until much later in human evolution. It will be outlined here and in the next section how many of the institutions that we associate with modern human life developed in two stages, one linked to the dispersal of modern humans from around 50,000 years ago, and one associated with the demographic expansion that occurred after the last glacial maximum (LGM) (<15,000 years ago).

After 50,000 years ago, humans, who had previously been confined to Africa and its adjacent western Asian landmass, spread very rapidly around the world. There is some evidence that along the Indian Ocean rim and into Australia this may have occurred earlier, prior to 60,000 years ago, but for more northerly parts of Eurasia a timeframe of between 50,000 and 30,000 is more appropriate (Lahr & Foley 1994). These multiple dispersals are associated with a number of new characteristics, which may be evidence for the emergence of novel social institutions. These include more regionally differentiated archaeological traditions, shorter timespans for the longevity of such traditions, greater evidence for the symbolic expression of individual and more probably ethnic identity, and special treatment of the dead in the form of burials (these last also occur among Neanderthals) (Klein 1992).

It may be suggested that these novel traits indicate certain types of social institution which did not exist before. As a whole they focus on one particular element, that of ethnic marking. No doubt some form of ethnicity would have been present in most hominin groups, for lineage-based patterns of residence and dispersal make the differentiation of communities a key element of all hominin populations. However, ethnic marking becomes very prominent in these populations. Cave art, personal ornamentation, and local differences in tool production are all interpretable in this way, and there is ample ethnographic evidence showing analogous situations. The social institution that seems to have developed is that of mechanisms to promote group cohesion through symbolic form. Two further institutional possibilities should be mentioned: the first is that this group cohesion may have been mediated through religious or quasi-religious practices, such as may be the case for ethnographically documented rock art in southern Africa (Lewis Williams 1981); the second is that the symbolic activity represents some form of ritually based activity which is differentiating members of the community, perhaps on the basis of age or sex — in other words, some form of initiation activity is occurring in which most probably men or just older men are given particular status. Whichever is the case, it can be argued that we see here evidence for the emergence of one or both of two social institutions — within-group roles representing power and/or status, or a mechanism for socially differentiating groups more markedly.

An important question to ask is: under what conditions does this emerge? These traits are not found universally. In many parts of the world, and at different times in the same parts of the world (for example, during the course of the European Upper Palaeolithic), these markers of social institutions either appear with greater or lesser intensity, or else they are completely absent. Such institutions, therefore, are not universal developmental stages in evolution (a somewhat tarnished evolutionary notion anyway), but are specific responses to local ecological conditions. It is certainly the case that the persistence of parietal art in France is associated with evidence for very dense human occupation in environments rich in both mammalian and aquatic resources. It can therefore be argued that what is happening in these particular regions is not part of a general evolutionary trend, but evidence for something we see consistently in later prehistory — complexity of social institutions arising as a response to high population densities. Other social institutions which it could be suggested may well be in place at this time are lineage-based descent communities, and the social mechanisms to establish stable social relationships, which may be either hierarchical or egalitarian; Woodburn (1982) has argued cogently that among hunter-gatherers both tendencies can be found, dependent upon local conditions.

Phase 2: the evolution of complex social institutions

What the Later Pleistocene (at most 100,000–20,000 years ago) shows is that the cognitive basis for emergent social institutions must have been present, but where such institutions do occur, they are patchily distributed, do not persist, and lack long-term trends. Collapse of such systems appears to have been an important element, and there is a lack of evidence for intensification. This contrasts markedly with the pattern found during the period after the last glacial maximum (<15,000 years ago). It is this contrast that will be addressed here.

During the period from 25,000 years ago to 15,000 years ago the climate deteriorated very markedly. The last glacial maximum was among the most intense periods of cold in recent earth history, and its effect on human populations was severe. There is extensive evidence to suggest that in many parts of the world, from Australia to Africa to Europe, there was a demographic contraction, and that in some places populations became extinct (Soffer & Gamble 1990). It can be suggested that it is this climatic event that underlies the lack of persistence in any trajectory towards permanent social complexity, and prevents the establishment of the emergent human social institutions discussed above prior to 15,000 years ago. In effect, it was perhaps the case that during the glacial maximum human social organization was more similar to that of the earlier populations of *H. sapiens* than it was to that of the period between 50,000 and 20,000 years ago, or at least that the event accounts for the very patchy distribution of any form of complexity. The hiatus in the development of human complexity over the later stages of the Pleistocene is strong evidence that ecologically sensitive factors are involved, not simply strictly biological or genetic ones, in the development of social institutions.

After 15,000 years ago, there was very rapid climatic warming, with a number of major effects (see Bar-Yosef, this volume). One of the most important of these was the spread of forests into areas of Europe previously dominated by mammal-rich steppes; another was the rise in sea level, destroying land bridges and greatly reducing important continental shelf zones such as those at the eastern end of the Mediterranean, in south-east Asia, and southern Africa; yet another was the climatic amelioration felt in parts of Europe. The response, seen in the archaeological record, is the growth of human populations across the world, and a return to a period of major human dispersals. While these can be seen in a number of areas, perhaps the most important spread zone was from the Middle East into northern Africa, southern Asia, and Europe (Renfrew 1987). This was a spread of agriculturalists. Similar but later dispersals occurred in eastern Asia and Africa. While these dispersals are associated with agriculture, in the Middle East at least this is not always the case. Bar-Yosef (this volume) has shown that they are integrally related to the expansion of the Natufian hunter-gatherer complex, so that it is difficult to disentangle the effects of a general post-Pleistocene growth in population from the spread of food-producing populations. It should also be noted that these hunter-gatherer dispersals also occurred in central Europe (Housley *et al.* 1997) and in Australia, where it is seen in the spread of Pama Nyungan languages (McGonvill 2001).

It is in this context that we see the first evidence for a number of emergent social institutions. These include areas associated with the performance of ritual (Bar-Yosef, this volume), evidence for organized religion at places such as Çatalhöyük (Mellaart 1964), evidence for fortifications and therefore inter-community violence, and possible organized warfare (Keeley 1995). The distribution of material culture across regions can also be said to indicate trading networks, and presumably a level of inter-community political organization. From this emerge, by the seventh millennium BC, the first signs of urban life, city-states, and the beginnings of larger state levels of organization, leading to the development of more clearly differentiated hierarchies and roles. More broadly across the world, there is evidence for interactions between very different types of societies — trade between boreal hunter-gatherers and sedentary agricultural societies in Europe, the spread of specialist nomadic pastoralists among others — as well as between major states.

The details of these developments are beyond the scope of this chapter. The key point, however, is that societies with evidence for far more permanent and diverse social institutions become increasing the norm during the Holocene, a process that can be said to continue with the rise and fall of many empires, and ultimately the establishment of an industrial process of production that has led to more and more incorporation of human populations and communities into a single system. The contrast with the Later Pleistocene is striking.

One interpretation might be that this is simply a lag effect — it took 40,000 years for the momentum of population growth to have the consequences that we see in the Holocene. However, as Richerson and Boyd (this volume) have shown, population growth or lack of it cannot account for such trends over such long-term timescales. A more probable explanation is that the hiatus between the appearance of modern humans and the full expression of social institutions is the result of climatic and ecological effects. As Eurasian populations grew during the Upper Palaeolithic, they developed levels of complexity and intensification that did produce novel social institutions. However, these occurred in areas in which hunter-gatherers dependent upon large mammals were prospering, during a time of partial glacial conditions. The last glacial maximum brought these to an abrupt end. The elaborate art of the Magdalenian, which is distributed across the LGM, shows that this system did persist in some areas. However, the post-glacial changes meant that the areas in which populations had the potential to grow and intensify were now located around the Mediterranean. Complexity emerged independently in these regions, but this time under very different ecological conditions, those of cereal-based agriculture. It was this system that had the potential for more permanent growth and expansion, and therefore it was the one in which the key element of social institutions — permanence within and between communities — could thrive.

In summary, the model presented here is one in which social institutions derive from the interaction of two components — biological capacity or propensity on the one hand, and ecological circumstances on the other. Each of these has varied over time, and has produced both a pattern which shows some level of progressive change — the trend across the Pleistocene at one time-scale, the development of civilizations in the Holocene at another — and a pattern which shows diversity, patchiness, and instability. In this sense the shape of the emergence of human social institutions is very similar to the shape of evolutionary developments more broadly — short-term diversity and local patterns, over which we can see a longer-term trend. More substantively, from the point of view of the central concern of this chapter, while some social institutions can be traced back over hundreds of thousands of years, the full expression of them did not occur with the origins of our species, but only when it became established in sufficiently dense and competitive communities.

THE EVOLUTION OF SOCIAL INSTITUTIONS

This attempt to construct a chronologically and ecologically sensitive model for the evolution of human social institutions can be brought together by considering some of the social institutions listed in Table 1.

Social communities

The tendency of humans to live in large groups can be considered to be the most basic building block of human social institutions, and to have an evolutionary history stretching back beyond the origins of the hominin lineage. This institution implies mechanisms by which individuals differentiate between those who belong to the community and those who do not, and behave accordingly.

Lineages and descent groups

It has been argued that the organization of many human communities into descent groups and lineages on the basis of kinship is one of the most fundamental elements of human social structure (Fortes & Evans-Pritchard 1940). The model developed here would strongly support this view. As human life history strategies became more extended, the potential for primate kin relationships to become inter-generational would have been realized. The development of descent groups and the central importance of the lineage as a social institution was almost certainly a gradual process over a long period of human evolution. The presence of this form of social structure would have led to other institutional developments, such as segmentary systems, kin-based fissioning of groups, and ultimately the development of such social phenomena as descent-based clans, tribes, and larger-scale political systems. These would have developed in relation to the increasing population density of the past 10,000 years or so, although they may also have occurred in some circumstances during the Later Palaeolithic.

Patrilocality and patrilineality

It has been argued that the ancestral condition for hominin communities was male residence and female dispersal, and as such hominin communities would have been male rather than female kin-bonded. If this was the case, then patrilocality would have been the norm for most of hominin evolution, and matrilocality or more flexible systems of community residence would be derived forms occurring under specific ecological conditions. The extension of patrilocality to patrilineality (and to other descent systems) would have derived from two phenomena. One of these would be the development of the cognitive capacity to transform residentially based relationships into more abstract kinship systems, and this cognitive state is likely to have been a phenomenon only of later human evolution, strongly linked to the evolution of language. The other would be the development of a more extended pattern of fission and fusion, so that, as discussed earlier, individuals in communities would retain notions of relationship and membership despite spatial location. It is the exploded pattern of fission–fusion that may have been one of the factors promoting the development of the cognitive systems underlying lineage systems.

Marriage

Although there is considerable variation from society to society, some form of marriage pattern is common to all. It was argued here that the transition to the *Homo ergaster* grade of evolution, associated with increased parental investment, more meat eating, and food sharing, may have promoted more exclusive patterns of mating than are found in chimpanzees or bonobos, and which may be inferred for the ancestral condition. As such, although marriage systems are most probably a later development, this pattern of exclusivity may have a more ancient origin. However, four points should be made in relation to this statement. First, the general process involved is more likely to have been the attachment of females to males, partly on account of the residential patterns, and partly because of the general principles of socioecology and comparative studies. Second, this exclusivity is unlikely to be specifically either monogamous or polygamous; ethnographic studies using socioecological principles have tended to see these systems as resource-sensitive, and so both are likely to have occurred, with variation within and between communities in actual behavioural patterns. Third, it is likely that the development of this institution would have occurred in conjunction with other mating and parenting strategies. There are reasons for seeing this phenomenon as related to parental investment, and mate choice (i.e. sexual activity) for both males and females may well have exhibited greater variation. Finally, marriage patterns and affiliative relationships are usually subsumed under the heading of kinship, but the model proposed here ascribes very different origins and antiquities to these two aspects of kinship. Male kin-bonding is the more ancient trait, and more exclusive mating the more derived one. Families are probably a more recent social institution than same-sex kin-bonded groups and larger social communities.

Institutions promoting social cohesion

A key element in the model developed here is that communities are fundamental. One probable consequence is that all such communities will have had social mechanisms for maintaining such groups, and excluding individuals that do not conform. These mechanisms, however, would have changed markedly over the course of hominin evolution, and perhaps one of the most striking phenomena of the period from 50,000 years ago is the increasing evidence for features which mark off one set of communities from others, or one community from another. The details are very obscure, but this would seem to be a key feature associated with the evolution of modern humans. It is likely that the archaeological record provides only a small glimpse into this behaviour, and language, social custom, dress, systems of cosmology, and coercive practices will be social institutions that would have developed in parallel. Once again, it is likely that ecological conditions would have shaped the intensity and the nature of these mechanisms, producing the wide variety in systems observed. It can be argued that religion, in two forms, would have been part of this process. The first of these is as a symbolic means of accounting for the relationships between members of the community and the wider human world — in other words, a cosmological function. The second, which probably developed more fully as societies became larger in scale, more differentiated, and more complex, would be as a system for maintaining social stability and a moral order.

Institutions relating to the internal social structure of communities

All social groups, human and non-human, have mechanisms for the maintenance of social relationships, from grooming to physical coercion. The main development that should be highlighted here is that, as populations grow and communities become larger and more sedentary, there is clearly a shift in balance from institutions that maintain egalitarianism, or at least reduce the growth of hierarchies, to those that codify such hierarchies. These probably arise with the greater potential for the control of resources that comes with small spatial scale, an absence of fission potential, and the development of economic systems in which storage is characteristic.

Institutions beyond the community

No community would have existed in isolation; comparative primate evidence shows that 'exogamy' pre-dates the origin of the hominin clade, and thus some form of relationship would always have existed between communities. An important component of these is likely to have been territorial exclusivity, certainly for males, and inter-group aggression and violence. The widespread distribution of warfare ethnographically is likely to have been a long-standing phenomenon (Keeley 1995), arising in general from the male kin-bonded nature of hominin communities. The growth of densely packed communities at the end of the Pleistocene, and perhaps in certain areas at other times, is likely to have greatly intensified this phenomenon, and led in many circumstances to situations where communities were predatory upon each other. However, it would also be the case that the same ecological circumstances would promote larger political structures that would bind together communities as well. These would be based upon social institutions derived from relationships between descent groups and patterns of marriage or female exchange. There would also be a growth of more economically oriented systems of integration.

CONCLUSION

Models of human evolution currently emphasize the importance of the origins of anatomically modern humans or *H. sapiens*. However, this chapter has shown that social institutions are no respecters of this boundary. Some social institutions, such as communities, male kin-bonding, exclusive mating patterns, and possibly descent groups were probably in existence long before the origins of our species; others may well have developed many thousands of years after the first *H. sapiens*. The origins of social institutions are therefore not located at a single point in time, but are scattered across our evolutionary history. It will be an enormous challenge to human evolutionary biology to unravel these many events and processes. Furthermore, such social institutions show evidence for ephemerality, in that they appear and disappear, or perhaps more precisely, vary in their expression and intensity in ways that are not simply directional. Directionality, that is the cumulative build-up of more and more complex institutions, does not appear to occur until the end of the Pleistocene, and is probably related to greater population densities and packing of communities, greater sedentism, and the development of food production. These three components are themselves interrelated.

If we are to be more specific about the origins of social institutions in time, then a number of key events might be provisionally identified.

1. The establishment of multi-male, multi-female communities operating a small-scale fission–fusion system, and held together by male residence and kinship. This probably occurred in the common ancestor of chimpanzees and hominins around 5 million years ago.

2. The development of a more expanded fission–fusion system, and hence spatially segregated communities held together by social relationships at a distance, and possibly involving more exclusive male–female bonds. This may have occurred shortly after 2 million years ago with *H. ergaster*.

3. The development of a greater capacity for language and symbolic thought, which would have transformed the way in which social institutions were maintained and changed, probably introducing greater variation in such systems. Prior to this, social institutions would have been maintained largely by direct physical mechanisms. This may well have occurred in the population that was ancestral to both Neanderthals and modern humans, which we would place at around 300,000 years ago (*H. helmei*), but others would either locate earlier or else apply specifically to the ancestors of modern humans alone.

4. The development of a greater need for communities to be ethnically identified, and perhaps for these both to be larger and to exist in a wider socially recognized network. This key event is particularly speculative, but may have occurred sporadically at least among early modern human populations to some extent, but developed extensively among later Pleistocene hunter-gatherer populations between 100,000 and 20,000 years ago.

5. The development of institutions related to complex inter-group relationships (both aggressive and co-operative), and to the maintenance of social and moral order in ways more likely to promote hierarchies. This development would have occurred sporadically among later Pleistocene hunter-gatherers in response to local ecological conditions, but is primarily associated with the changes occurring at the end of the Pleistocene. It was this development, rather than the evolution of modern humans, which set in train the massive rise in social complexity that has occurred in the past few millennia.

Although much of the model with the inferences developed here must be treated with great caution, since the nature of the evidence available is sparse indeed, none the less the general principle underlying this reconstruction should be emphasized. Social institutions neither arise from the innate propensities of the human species, of which there are many, nor come solely from cultural responses to the socioecological circumstances in which populations find themselves. They arise from the interaction between the two. Both elements have changed over the course of history and prehistory, and the latter at least continues to change into the present day. This interaction has given rise both to strong directional change — clear trajectories that in another and more innocent age might have been called progressive — and also to smaller fluctuations that add complexity to the pattern.

REFERENCES

AIELLO, L.C. & WHEELER, P. 1995: The expensive tissue hypothesis. *Current Anthropology* 36, 199–222.

BARKOW, L., COSMIDES, L. & TOOBY, J. (eds.) 1992: *The Adapted Mind* (Oxford, Oxford University Press).

BETZIG, L., BORGERHOFF-MULDER, M. & TURKE, P. (eds.) 1988: *Human Reproductive Behaviour: a Darwinian Perspective* (Cambridge, Cambridge University Press).

BORGERHOFF MULDER, M. 1996: Responses to environmental novelty: changes in men's marriage strategies in a rural Kenyan community. *Proceedings of the British Academy* 88, 203–22.

BROOKS A.S., HELGREN, D.M., CRAMER, J.S., FRANKLIN, A., HORNYAK, W., KEATING, J.M., KLEIN, R.G., RINK, W.J., SCHWARCZ, H.P., SMITH, J.N.L., STEWART, K., TODD, N.E., VERNIERS, J. & YELLEN, J.E. 1995: Dating and context of three Middle Stone Age sites with bone points in the Upper Semliki Valley, Zaire. *Science* 268, 548–53.

DEACON, H.J. 1989: Late Pleistocene palaeoecology and archaeology in the southern Cape, South Africa. In Mellars, P. & Stringer C.B. (eds.), *The Human Revolution* (Edinburgh, Edinburgh University Press), 547–64.

DEACON, H.J. & SHUURMAN, R. 1992: The origins of modern people: the evidence from Klasies River. In Brauer, G. & Smith, F.H. (eds.), *Continuity or Replacement? Controversies in* Homo sapiens *Evolution* (Rotterdam, Balkema), 121–9.

DEAN, M.C., STRINGER, C.B. & BROMAGE, T.G. 1986: Age at death of the Neanderthal child from Devil's Tower, Gibraltar and implications for the study of general growth and development in Neanderthals. *American Journal of Physical Anthropology* 70, 301–9.

FOLEY, R.A. 1987: *Another Unique Species: Patterns in Human Evolutionary Ecology* (London, Longman).

FOLEY, R.A. 1989: The evolution of hominid social behaviour. In Standen, V. & Foley, R. (eds.), *Comparative Socioecology: the Behavioural Ecology of Humans and Other Mammals* (Oxford, Blackwell Scientific Publications), 474–93.

FOLEY, R.A. 1994: The evolution and adaptive significance of hominid maternal behaviour. In Pryce, C.R., Martin, R.D. & Skuse, D. (eds.), *Motherhood in Human and Nonhuman Primates* (Bas, Karger), 27–36.

FOLEY, R.A. 1995: *Humans before Humanity* (Oxford, Blackwell).

FOLEY, R.A. & LAHR, M.M. 1997: Mode 3 technologies and the evolution of modern humans. *Cambridge Archaeological Journal* 7, 3–36.

FOLEY, R.A. & LAHR, M.M. in press: The anthropological, demographic and ecological context of human evolutionary genetics. In Donnelly, P. & Foley, R.A. (eds.), *Genes, Fossils and Behaviour: an Integrated Approach to Human Evolution* (Brussels, IOS Press).

FOLEY, R.A. & LEE, P.C. 1989: Finite social space, evolutionary pathways and reconstructing hominid behaviour. *Science* 243, 901–6.

FOLEY, R.A. & LEE, P.C. 1991: Ecology and energetics of encephalization in hominid evolution. *Philosophical Transactions of the Royal Society*, London Series B 334, 223–32.

FOLEY, R.A. & LEE, P.C. 1995: Finite social space and the evolution of human social behaviour. In Steele, J. & Shennan, S. (eds.), *The Archaeology of Human Ancestry* (London, Routledge), 47–66.

FORTES, M. & EVANS-PRITCHARD, E. 1940: *African Political Systems* (Oxford, Oxford University Press).

GHIGLIERI, M.P. 1987: Sociobiology of the great apes and the hominid ancestor. *Journal of Human Evolution* 16, 319–57.

GOODALL, J. 1986: *The Chimpanzees of Gombe* (Cambridge, MA, Harvard University Press).

HAMMER, M. & FOLEY, R.A. 1996: Longevity and life history in hominid evolution. *Human Evolution* 11, 61–6.

HAYDEN, B. 1981: Subsistence and ecological adaptation of modern hunter-gatherers. In Harding, R.S.O. & Teleki, G. (eds.), *Omnivorous Primates* (New York, Columbia University Press).

HINDE, R.A. 1987: *Individuals, Relationships and Culture: Links Between Ethology and the Social Sciences* (Cambridge, Cambridge University Press).

HOUSLEY, R.A., GAMBLE, C.S., STREET, M. & PETTITT, P. 1997: Radiocarbon evidence for the late glacial human recolonization of northern Europe. *Proceedings of the Prehistoric Society* 63, 25–54.

ISAAC, G.L., 1978: The food-sharing behavior of protohuman hominids. *Scientific American* 238, 90–108.

KAPLAN, H., HILL, K., LANCASTER, J. & HURTADO, A.M. 2000: A theory of human life history evolution: diet, intelligence and longevity, *Evolutionary Anthropology* 9, 156–85.

KEELEY, L. 1995: *War before Civilization* (Oxford, Oxford University Press).

KELLY, R. 1995: *The Foraging Spectrum* (Washington, DC, Smithsonian Institution Press).

KLEIN, R.G. 1992: The archaeology of modern human origins. *Evolutionary Anthropology* 1, 5–15.

KLEIN, R.G. 1995: Anatomy, behaviour, and modern human origins. *Journal of World Prehistory* 9, 167–98.

KLEIN, R.G. 1999: *The Human Career* (Chicago, Chicago University Press).

KLEIN, R.G. 2000: Archaeology and the evolution of human behaviour. *Evolutionary Anthropology*, 9 (1), 17–36.

KLEIN, R.G. & CRUZ-URIBE, K. 1996: Exploitation of large bovids and seals at Middle and Later Stone Age sites in South Africa. *Journal of Human Evolution* 31, 315–34.

KREBS, J. & DAVIES, N. 1997: *Behavioural Ecology: an Evolutionary Approach*, 4th edn (Oxford, Blackwell).

LAHR, M.M. & FOLEY, R.A. 1994: Multiple dispersals and the origins of modern humans. *Evolutionary Anthropology* 3(2), 48–60.

LAYTON, R.H., FOLEY, R.A. & WILLIAMS, E. 1991: The transition between hunting and gathering and specialized husbandry of resources: a socioecological approach. *Current Anthropology* 32 (3), 255–74.

LEE, P.C. (ed.) 1999: *Comparative Primate Socioecology* (Cambridge, Cambridge University Press).

LEE, R.B. & DEVORE, I. 1968: *Man the Hunter* (Chicago, Aldine).

LEWIS WILLIAMS, J.D. 1981: *Believing and Seeing: Symbolic Meanings in Southern San Rock Art* (London, Academic Press).

LOVEJOY, C.O. 1981: The origin of man. *Science* 211, 341–50.

McCONVELL, P. 2001: Language shift and language spread among hunter-gatherers. In Panter-Brick, C., Leyton, R.H. & Rowley-Conway, P. (eds.), *Hunter-Gatherers: an Interdisciplinary Perspective* (Cambridge, Cambridge University Press), 143–69.

McLARNON, A. 1996: The evolution of the spinal cord in primates: evidence from the foramen magnum and the vertebral canal. *Journal of Human Evolution* 30, 121–38.

MELLAART, J. 1964: Excavations at Çatalhöyük. *Anatolian Studies* 14.

MELLARS, P. 1989: Technological changes at the Middle–Upper Palaeolithic transition: economic, social and cognitive perspectives. In Mellars, P. & Stringer, C.B. (eds.), *The Human Revolution* (Edinburgh, Edinburgh University Press), 338–65.

MELLARS, P. 1996: *The Neanderthal Legacy: an Archaeological Perspective from Western Europe* (Princeton, Princeton University Press).

NETTLE, D. 1999: *Linguistic Diversity* (Oxford, Oxford University Press).

O'CONNELL, J.F., HAWKES, K. & JONES, N.G.B. 1999: Grandmothering and the evolution of *Homo erectus. Journal of Human Evolution* 36, 461–85.

RENFREW, C. 1987: *Archaeology and Language: the Puzzle of the Indo-European Origins* (London, Penguin Books).

RODSPETH, L., WRANGHAM, R.W., HARRIGAN, A.M. & SMUTS, B.B. 1991: The human community as a primate society. *Current Anthropology* 32, 221–54.

RUNCIMAN, W.G. 1998. *The Social Animal* (London, Harper Collins).

SMITH, B.H. 1989: Dental development as a measure of life history in primates. *Evolution* 43, 683–8.

SMITH, B.H. 1992: Life history and the evolution of human maturation. *Evolutionary Anthropology* 1 (4), 134–42.

SMITH, E.A. & WINTERHALDER, B. (eds.) 1992: *Evolutionary Ecology and Human Behavior* (Chicago, Aldine de Gruyter).

SOFFER, O. & GAMBLE, C. (eds.) 1990: *The World at 18,000 BP, Vol. I* (London, Unwin Hyman).

STANDEN, V. & FOLEY, R.A. (eds.) 1989: *Comparative Socioecology: the Behavioural Ecology of Humans and Other Mammals* (Oxford, Blackwell Scientific Publications).

STANFORD, C. 1999: *The Hunting Apes* (Princeton, Princeton University Press).

STINER, M.C., MUNRO, N.D., SUROVELL, T.A., TCHERNOV, E. & BAR-YOSEF, O. 1999: Paleolithic population growth pulses evidenced by small animal exploitation. *Science* 283, 190–4.

WHITE, T.D. 1987: Cannibals at Klasies? *Sagittarius* 4, 6–9.

WINICK, C. 1960: *Dictionary of Anthropology* (London, Peter Owen).

WOODBURN, J. 1982: Egalitarian societies. *Man* 17, 431–51.

WRANGHAM, R.W. 1980: An ecological model of female-bonded primate groups. *Behavior* 75, 262–99.

WRANGHAM, R.W. 1986: Ecology and social relationships of two species of chimpanzee. In Rubenstein, D.I. and Wrangham R.W. (eds.), *Ecological Aspects of Social Relationships in Birds and Mammals* (Princeton, Princeton University Press), 352–78.

WRANGHAM, R.W. 1987: The significance of African apes for reconstructing human evolution. In Kinzey, W.G. (ed.), *The Evolution of Human Behavior: Primate Models* (Albany, SUNY Press), 28–47.

WRANGHAM, R. & PETERSON, D. 1996: *Demonic Males* (London, Bloomsbury).

Institutional Evolution in the Holocene: The Rise of Complex Societies

PETER J. RICHERSON
&
ROBERT BOYD

INTRODUCTION

HUMAN SOCIETIES ARE MUCH LARGER and more complex than the societies of other social mammals. This fact creates an evolutionary puzzle. Five million or so years ago, our ancestors lived in small groups with limited co-operation organized around kinship and reciprocity. Today, we live in vast societies, organized and regulated by many complex institutions. In this chapter, we argue that this transition occurred in two stages. First, over the past several hundred thousand years, humans evolved the capacity for cumulative cultural evolution, which, in turn, led to the gene-culture co-evolution of larger and more co-operative societies. By the Late Pleistocene, hominids evolved the social instincts necessary to create societies on the tribal scale, a level of sociocultural organization absent in other primates and, indeed, entirely unique to our species. These instincts and the institutions that they underpinned were the preadaptations to complex sociality that followed. Second, the Pleistocene–Holocene boundary, about 11,500 years ago, marks a major transition point in human social evolution. Institutional evolution in the Late Pleistocene was limited by a regime of highly variable environments under which agricultural subsistence systems were *impossible*. The climate of the Holocene has been very much less variable, and agriculture is *possible* over a large fraction of the earth's land surface. Indeed, the greater efficiency of agricultural production means that agricultural populations can generally out-compete hunter-gather populations. Thus once agriculture became possible, competitive forces made it *compulsory*, in the long run at least. We hypothesize that a similar dynamic drove the evolution of social institutions. Societies with more co-operation, co-ordination, and division of labour can generally out-compete societies with less.

Proceedings of the British Academy, **110**, 197–234, © The British Academy 2001.

Because the Pleistocene–Holocene transition was a rapid, globally synchronous event, variations in the rate of institutional evolution in different parts of the world represent natural experiments that should yield clues pointing to the processes that limit the rate of evolution of institutions. That is, since the progressive trend towards more complex societies characterizes almost all parts of the world, we know that the equilibrium degree of complexity has not been reached until quite recently at least. (We make no attempt to speculate about such questions as how much more complex societies can become or whether industrial use of non-renewable resources has created an unsustainable overshoot of equilibrium.) Thus we can conceive of the problem as discovering the main limiting factors that slow the competition-driven progressive trend towards greater social complexity. A number of plausible candidates exist, permitting a dim outline of the large-scale dynamics of institutional evolution.

Darwinian models of cultural evolution

Two rather different approaches to the use of Darwinian theory are current in the contemporary social sciences. One is to apply the substantive results of Darwinian theory to human behaviour. This field was pioneered by Alexander (1974) and Wilson (1975) and was given a somewhat different twist by Symons (1989) under the heading of 'evolutionary psychology'. Since natural selection is the most important directional force in organic evolution, these scholars use fitness optimizing models to generate testable hypotheses about human behaviour (Borgerhoff Mulder *et al.* 1997). Typically, such work endorses a number of common dogmas current in evolutionary biology, for example the generalization that group selection is seldom a strong force. The weakness of this approach is that it may not do full justice to the unique features of human behaviour.

We advocate a different strategy pioneered by Campbell (1965) and first put in mathematical form by Cavalli-Sforza and Feldman (1973). This work starts with the idea that culture is a system of inheritance. We acquire culture by imitating other individuals, much as we get our genes from our parents. The existence of a fancy capacity for high-fidelity imitation is one of the most important derived characters distinguishing us from our primate relatives, who have only relatively rudimentary imitative abilities (Tomasello 1999). We are also an unusually docile animal (Simon 1990) and unusually sensitive to expressions of approval and disapproval by parents and others (Baum 1994: 218–19). Thus parents, teachers, and peers can shape our behaviour rapidly and easily compared with training other animals using more expensive material rewards and punishments. Finally, once children acquire language, parents and others can communicate new ideas quite economically to those who don't know them. This economy is only relative; although we get our genes all at once at the moment of conception, acquiring an adult cultural repertoire takes some two decades. Humans ultimately acquire a repertoire of culture that rivals the genome in size.

The existence of cultural transmission means that culture has what evolutionary biologists call 'population level properties'. Individuals' behaviour depends on the behaviours common in the population from whom they acquire beliefs, just as individuals' anatomy is dependent on the genes common in the population from whom they acquired their genes. The diversity of cultural traits across cultures is great, but for the most part we are limited to learning those extant in our culture in our time. However, in the long run, the commonness or rarity of genes or culture in the population is a product of what happens to the individuals who reproduce or not, and are imitated or not. The analogy is more than a curiosity because population biologists have developed a formidable kit of empirical and theoretical tools to analyse this intricate interplay between the individual and population level. In the terms sociologists often use, population biologists have the means to make the macro–micro problem (Alexander *et al.* 1987) tractable. Several theorists, but fewer empiricists, have raided the population biologists' cupboard for these tools (Boyd & Richerson 1985; Cavalli-Sforza & Feldman 1981; Durham 1991; Lumsden & Wilson 1981).

In this exercise, we think it best to wear the analogy between genes and memes most lightly. For example, we have resisted using the term 'meme' to describe the 'unit' of cultural transmission (Boyd & Richerson 2000). Who knows whether the structure of cultural inheritance is anything like the neatly particulate gene? We do know that culture is most un-gene-like in many respects. Culture has the principle of inheritance of acquired variation (what one person invents, another can imitate). We are not entirely blind victims of chance imitation, but can pick and choose among any cultural variants that come to our attention and creatively put our own twist on them. We don't have to imitate our parents or any other specific individuals, but can always be open to a better idea. The innovative part of the Darwinian analysis of cultural evolution has been to explore the impact of such differences on the cultural evolutionary process, letting model results and empirical facts, not substantive analogies, guide the research. Substantively, cultural evolution turns out to have its own unique adaptive properties and its own unique suite of characteristic maladaptations, some examples of which we discuss here.

Maladaptations are epistemologically more interesting than adaptations. The trouble with adaptations is that the competing theories — creationism, genetic fitness optimizing, cultural evolution, macrofunctionalism, rational choice theory — all predict that adaptive behaviour will be common. Each theory's predicted maladaptations are much more distinctive. For example,

Hamilton (1964) deduced from the principles of natural selection acting on genes that organisms should engage in altruistic acts only when the benefit to the recipient exceeds the costs to the provider by a factor greater than the reciprocal of the relatedness by common descent between them, his famous $b/c > 1/r$ rule. Since in most animal species individuals have only few relatives with appreciable r, Hamilton's theory predicts that altruism will be massively undersupplied compared with a perfectly group-selected case where altruism within groups should be supplied whenever $b/c > 1$. Every individual in a group would be better off if every other individual followed the $b/c > 1$ rule instead of the $b/c > 1/r$, but natural selection on genes cannot favour such acts. With the exception of humans and a few other special cases, Hamilton's rule predicts the maladaptively low amount of animal co-operation quite well. Human societies are a theoretical puzzle because they typically include much co-operation between distantly related and unrelated people. We have adaptively evaded a rule that otherwise seems to have almost the law-like force of a physical principle. We argue below that cultural evolution is likely to be the source of our capacity to pull off our defiance of Hamilton's rule.

The unique features of the cultural system of inheritance are predictable from the elementary consideration that selection on genes to increase our capacity to learn from each other would surely not have favoured this rather costly system if it did only what genes could do for themselves. One important advantage of the cultural system is the linkage of decision-making processes with transmission to create a system for the inheritance of acquired variation. Given that decision rules ultimately derive from the action of selection on genes and hence are adaptive, on average at least, a system that responds both directly to natural selection *and* to adaptive decision-making forces will be able to adapt to varying environments more quickly than can organisms that adapt by genes and non-transmitted learning (Boyd & Richerson 1985: Chapters 4 and 5). Plagiarizing the learning of others creates a system that can adapt swiftly to new conditions without a crippling expenditure of effort on individual learning. Second, accurate and rapid social learning allows humans, but seemingly not other species, to accumulate innovations so as to build up, historically over many generations — but rather rapidly compared with organic evolution — more sophisticated cultural adaptations than individual people could possibly have invented for themselves (Boyd & Richerson 1996). Human cultural adaptations are not only dramatically different from place to place and time to time, but are also as complex as organic adaptations that would take much longer to evolve. The Inuit adaptation to the Arctic and the San adaptation to the Kalahari are impressively complex and impressively different on a scale that would result in different species if accomplished by organic evolution. In support of these theory-derived conjectures, we note that humans evolved during the Pleistocene, a period of high-frequency climatic variation (Richerson &

Boyd 2000), and we became an unusually widespread animal by Middle Pleistocene times. The ability to adapt quickly to a temporarily variable environment is easily put to use adapting to spatial variation as well, permitting a tropical ape to live in temperate and eventually periglacial climates. We became completely cosmopolitan using subsistence strategies tailored to practically every terrestrial and amphibious habitat on the planet. We believe that ability of the cultural system to rapidly create sophisticated adaptations to niches that persisted for a relatively few generations was the main advantage that paid the overhead of our large brain and long learning curve.

The evolution of institutions of complex societies

The evolution of complex societies is one of the most interesting questions in all the social sciences. How can a species long adapted to living in small egalitarian groups evolve revolutionary new social institutions that lead them to live in very large, highly inegalitarian social systems? Tribal people often express shock and contempt at what we put up with in the name of 'civilization'. Why did the progressive trajectory of increased complexity start around 10,000 years ago, not 30,000 or 5,000? Why did societies in some parts of the world move down the progressive path more swiftly than others? What processes regulate the tempo of institutional evolution? What gives the progressive trend its multilinear diversity? No two trajectories of complexification are identical, even in closely related societies and sub-societies, much less in remotely connected cases such as western Europe, western Asia, India, China, and Mesoamerica, despite many similarities. Why has the pace of change had a tendency to accelerate as we approach the present? Why is the progressive trend punctuated, in every historical case, by more or less abrupt declines and collapses?

These are exceedingly complex questions that have defied definitive solution despite much hard work — and much real progress — by social scientists, historians, and political philosophers. The development of Darwinian tools encourages a fresh cut at them. In what follows we lay out an analysis that seems like a sensible series of first steps using Darwinian analysis of Holocene institutional evolution. We beg our readers' indulgence with the inevitable crudities that accompany first steps and hope that you will take them as illustrations prefiguring what a more mature analysis will most likely accomplish. The boast of Darwinian biologists is that the power of their theory in that discipline derives first, from its correct conception of the processes of evolution, and second, from its inclusive, synthetic, and systemic commitments. Darwinian biology is a big tent housing diverse and often fractious practitioners. Even after a century and a half of work it is a vibrant field full of interesting unsolved puzzles, many of a quite fundamental character.

We hope that this chapter conveys some vision, however limited it is in the present state of development of our field, of what the social sciences might look like if the use of Darwinian methods became routine in the analysis of culturally determined behaviours.

TRIBAL SOCIAL INSTINCTS HYPOTHESIS

The tribal social instincts hypothesis is based on the belief that group selection plays a more important role in shaping culturally transmitted variation than it does in shaping genetic variation, and, as a result, that humans have lived in social environments characterized by high levels of co-operation for as long as culture has played an important role in human development. The simplest model of group selection on cultural variation we have made is based on the effects of a conformist bias in cultural transmission (Boyd & Richerson, 1985: Chapter 7; Henrich & Boyd 1998). Conformity is a useful rule to follow in imitating others because many evolutionary forces conspire to make adaptive behaviour common. When in doubt, doing as the Romans do when in Rome is an easy and useful rule to follow. Using this rule has the effect of reducing variation within groups and protecting groups against the effects of migration from other groups. Other rules, such as preferring to imitate people of your own symbolically marked group or the practice of social selection against deviants, may have similar effects. Group selection does not work on genes for co-operation according to most models because group selection cannot easily build variation between groups as fast as selection against co-operators within groups — and migration between groups — reduces it. Thus, selection on cultural variation is a more likely mechanism for favouring the origins of co-operative institutions than is selection on genes. We have also studied models of the evolution of symbolic marking of group boundaries (Boyd & Richerson 1987; McElreath *et al.* no date) and moralistic punishment (Boyd & Richerson 1992a).

By the Late Pleistocene, 50,000 BP, perhaps earlier, human societies probably possessed tribal-scale institutions (Bettinger 1991; Richerson & Boyd 1998). If we define 'institutions' as customary rules of behaviour that have the effect of creating sociopolitical structures serving collective functions, then hunting and gathering societies of the ethnographic record always have tribal-scale institutions, though sometimes rather minimal ones. For example, many of the simplest known hunting societies have well-developed systems of egalitarian counter-dominance that prevent individuals from appropriating disproportionate shares of food. These institutions in turn probably allow such societies to act as effective risk-sharing groups that can efficiently exploit high return, high-risk strategies such as the pursuit of big game (Boehm 1993; Wiessner 1996). Ethnographically known hunter-gatherers are quite variable in the scale, sophistication, and formality of their institutions (Arnold 1996; Kelly 1995).

The Shoshoni of the American Great Basin are a classic example of socially very simple hunter-gatherers. Steward (1955: Chapter 6) described them as having a family level of sociocultural integration. During most of the year, Shoshoni nuclear families foraged for sparse plant resources alone or in the company of one or two other families. The main resources of the Great Basin did not favour co-operation in subsistence activities and the low productivity of the environment discouraged concentrations of population favourable to other social activities. The Shoshoni lacked a formal tribal political system and had no organized religious activities. Nevertheless, as Steward emphasized, even the Shoshoni had *some* customs regulating social life and *some* routine collective behaviour. The kinship system itself regulated social relations between families. Marriage was in the form of a contract between the families involved. A common property system ensured that all families had equal access, first come, first served, to most but not all resources. During the winter, families aggregated into multi-family camps. In such camps *ad hoc* 'bosses' organized events such as communal rabbit and antelope hunts, dances, and games. Respected men served as regional repositories of information about the distribution of subsistence resources and so regulated the dispersion and assembly of families during the seasons.

At the opposite extreme, some hunting and gathering societies had much denser populations and much more extensive and formal economic, political, and religious institutions than the Shoshoni. In western North America, the salmon-rich societies of the north-west coast are such an example (Kelly 1995: 321–8). Many of these societies were ranked and chiefs had much power. Politics, religion, and the economy were highly organized. Between these extremes a great diversity of institutional forms has been recorded among hunter-gatherers. In most of California, for example, people were organized into 'tribelets' comprising anywhere from a hundred to a few thousand people (Bean 1978). Most ethno-linguistic units were divided into several autonomous tribelets. Tribelets were generally composed of corporate lineages and had formal political leadership, sometimes including ranked chieftainships. Tribelets were generally centred on a principal village where council meetings, religious rituals, and collective economic activities took place. Supra-tribelet institutions included regional cult complexes and trade fairs. These brought thousands of people together for annual or more frequent gatherings, often including people from different ethno-linguistic groups.

Did the range of Late Pleistocene hunting and gathering societies resemble the ethnographic range? Extrapolating from ethnographic to archaeological cases is of course fraught with problems, especially in the case of hunter-gatherers most of whose material culture, for example dwellings, is poorly preserved (Bettinger 1991; Kelly 1995: Chapter 9). Archaeology is relatively silent about social organization but a good case can be made that at least Late Pleistocene societies were towards the complex end of the ethnographic spectrum (Price & Brown 1985).The hunting of big game is a subsistence strategy that generally involves co-operation in hunting, fairly large-scale risk-sharing social strategies, and hence social institutions considerably more complex than the Shoshonean extreme. Big game hunting was common in the Late Pleistocene and almost certainly favoured the same relatively complex institutions as it does in ethnographic cases. The personal art that is a conspicuous part of the Upper Palaeolithic transition is similar to craft productions that are incorporated into tribal-scale institutions in ethnographic cases (Wiessner 1984). The cave art of France and Spain is the sort of activity associated with fairly large-scale ritual systems in ethnographic cases. Insofar as the archaeological record reflects social institutions it suggests that Late Pleistocene societies had institutions on the tribal scale comparable to those observed by ethnographers.

We believe that the human capacity to live in tribes evolved by the co-evolution of genes and culture. Simple cultural co-operative institutions favoured by cultural group selection would have favoured genotypes that were better able to live in groups that at first were only marginally co-operative outside of families and simple schemes of reciprocity. Given marginal genetic changes, cultural evolution could marginally advance the scale of co-operation. These rounds of co-evolutionary change then proceeded until capacities for co-operation with distantly related fellow tribals, emotional attachments to symbolically marked groups, and willingness to punish others for transgression of group rules became quite advanced. One doesn't have to search far for mechanisms by which cultural institutions might exert forces tugging in this direction. Cultural norms affect mate choice and people seeking mates are likely to discriminate against genotypes that are incapable of conforming to cultural norms (Richerson & Boyd 1989). Men who cannot control their self-serving aggression end up exiled to the wilderness in small-scale societies and to prison in contemporary ones. Women who are lazy or an embarrassment in social circumstances are unlikely to find or keep husbands. We believe that with, at a minimum, tens of thousands of years to work with, natural section on cultural variation could easily have had dramatic effects on the evolution of human genes by this process. Of course, humans are still in part a wild animal; our genetically transmitted evolved psychology shapes human cultures, and as a result cultural adaptations often still serve the ancient imperatives of genetic fitness. But the leash works both ways. Cultural evolution creates new selective environments that cause *cultural imperatives to be built into our genes.*

Almost everyone agrees that human cultures were essentially modern by the

Upper Palaeolithic, 50,000 years ago. So even if the cultural group selection process began as late as the Upper Palaeolithic, human behaviour has been selected for 2,000 generations in social environments in which the innate willingness to recognize, aid, and if necessary, punish fellow group members was favoured by social selection acting on genes. We suppose that the resulting tribal instincts are something like principles in the Chomskian linguists' 'principles and parameters' view of language (Pinker 1994). The innate principles furnish people with basic predispositions, emotional capacities, and social skills that are implemented in practice through highly variable cultural institutions, the parameters. People are innately prepared to act as members of tribes, but culture tells us how to recognize who belongs to our tribes, what schedules of aid, praise, and punishment are due to tribal fellows, and how the tribe is to deal with other tribes — allies, enemies, and clients. Richerson & Boyd (in press) review the empirical evidence supporting the tribal social instincts hypothesis.

Because the tribal instincts are of relatively recent origin, they are not the sole regulators of human social life. The tribal instincts are laid on top of more ancient social instincts rooted in kin selection and reciprocal altruism. These ancient social instincts conflict with the tribal. We are simultaneously committed to tribes, family, and self, even though the conflicting demands very often cause us great anguish such as Freud (1930) described in *Civilization and Its Discontents* or Graham Greene portrayed in novels such as *The Honorary Consul*. The existence of ancient instincts significantly constrains the evolution of institutions.

Competing hypotheses

We have not the space to review in detail all the competing hypotheses to explain the evolution of human social organization. Broadly speaking, however, these fall into two classes: those that emphasize individual level processes and those that emphasize group functionality. Methodological individualists in the social sciences are deeply sceptical about the group-functional picture of human behaviour, and wish to ground the social sciences on the postulate of self-interested rational choice (e.g. Coleman 1990). Evolutionary biologists by and large follow Williams' (1966) lead in rejecting group selection as an important force in nature. In the case of humans, not to mention other animals, selfish behaviour and very small-scale altruism, for example among close relatives, are common and in accord with methodological individualists' theoretical models. Following Axelrod and Hamilton (1981) and Alexander (1987), individualists reckon that the logic of small-scale reciprocity can be scaled up to explain human co-operation on the large scale without violating any of the standard assumptions of methodological individualism, such as postulating a strong role for group selection.

The relationship between rational choice theory and cultural evolution theory is complex because we assume that individual choice exists and acts as a force shaping cultural evolution (Boyd & Richerson, 1993). We hold, however, that choice is marginal and does not normally follow the canons of formal rationality. That is, people form their repertoires of behaviour mostly by imitation of others, making somewhat biased choices among the cultural variants they observe, and sometimes independently inventing new adaptive behaviours. When the results of such myopic decision-making are accumulated over a population of people and many cycles of imitation and decision-making they indeed become potent evolutionary forces. However, we also suppose that they are not sufficiently powerful to obviate the effects of natural selection *on cultural variation*. Our social instincts hypothesis requires that cultural group selection be strong enough to counter individualistically motivated selfish decision-making in order to favour tribal-scale co-operation. Then, once a social instinct favouring pro-social behaviour towards ingroup members exists, it will affect not only everyday behaviour, but also the kinds of decisions people make about what new cultural variants to adopt. People will tend to bias their 'vote' in favour of new pro-social institutions, as indeed seems to be the case in American voting patterns (Sears & Funk 1990). Humans seem to be moved by both selfish and pro-social arguments. Which wins, and to what extent, in any given case is problematical. Our ancestors, lacking the tribal social instinct, had not even our ambivalent commitment to ingroup co-operation and in the end our societies come to differ dramatically from theirs.

Group functionalism was once very prominent in sociology and anthropology. Most functionalist hypotheses have been silent about evolutionary origins and so are not of interest to us here. Several evolutionary hypotheses have been proposed since Darwin (1874: 179) articulated a clear group selection argument to account for human co-operation: 'A tribe including many members who, from possessing in a high degree the spirit of patriotism, fidelity, obedience, courage, and sympathy, were always ready to aid one another, and to sacrifice themselves for the common good, would be victorious over most other tribes; and this would be natural selection.' One possibility is that humans are genetically group selected. Several prominent modern Darwinians (Alexander (1974), Eibl-Eibesfeldt (1982) Hamilton (1975), and Wilson (1975: 561–2)) have given serious consideration to group selection as a force *in the special case* of human ultra-sociality. They are impressed, as we are, by the organization of human populations into units that engage in highly organized, lethal competition with other groups, not to mention other forms of co-operation. Direct group selection on genes is a process that could give human groups a degree of functional integration. A second view is that processes peculiar to culture are prone to group selection. This idea is the root of our tribal instincts hypothesis.

A third possibility is that human propensities to co-operate are a byproduct or accident of some other process. Simon (1990) proposed that human co-operation is a byproduct of our docility and that docility is necessary to take advantage of cultural transmission. We worry that this hypothesis can be true at the margin. Selfish and manipulative individuals do not seem to be automatically handicapped in their acquistion of culture. Van den Berghe (1981) argued that in small-scale societies cultural similarity in dialect, clothing, and so forth was used as a sensitive marker of genetic relatedness. The relative isolation of families and bands set up sharp cultural gradients that would measure genetic distance more effectively than innate characters for which the gradient at the small scale is likely to be very small. In the much larger, denser societies made possible by agriculture, the number of people with very similar culture might reach thousands and, with mass media, millions. Such cultural similarity may trigger kin-selected social instincts so that we treat our fellow tribals as close kin. The problem with this hypothesis is that in many circumstances we still recognize our kin and behave as if our innate propensity to favour real kin is quite intact. For example, blood relatives tend to be spared homicide relative to non-kin, such as stepchildren in the same household (Daly & Wilson 1988).

WORK-AROUND HYPOTHESIS

Contemporary human societies differ drastically from those under which our social instincts presumably evolved. Until a few thousand years ago, humans lived in relatively small, egalitarian societies with a modest division of labour. After the domestication of plants and animals, beginning about 11,500 years ago, human densities rose substantially and the potential for an expanded division of labour grew. Beginning about 5,000 years ago, complex societies began to emerge. Hierarchical states arose to administer the increasingly minute division of labour. Families became dependent on the products of strangers for routine subsistence. Leaders came to have great and sometimes quite arbitrary power to coerce common citizens. Complex systems also universally develop social stratification in which objective material well-being and culturally defined prestige vary greatly by social role. Those in high positions in the command and control system seemingly inevitably acquire a more or less disproportionate share of society's rewards. There is every evidence that humans' Pleistocene evolutionary experience did not prepare us to tolerate more than the most minimal command and control institutions (Boehm 1993). Nor were we prepared to tolerate much inequality. The cultural evolution of complex societies in the Holocene will have had to *work around* these awkward realities of our ancient and tribal instincts, drawing upon the pro-social elements in them while finessing the elements not suited to large-scale social systems.

If our tribal social instincts hypothesis is correct, complex societies will have evolved under the constraints and possibilities offered by our evolved social psychology (Salter 1995). The rapid social changes of the past few thousand years should throw our social instincts into high relief. For example, one of the most striking features of complex societies, including modern societies, is the persistence of tribal-scale social institutions and the elaboration of institutions such as nationalism that utilize mass media to simulate tribes on a larger scale. Business organizations, schools, religions, and government bureaucracies generally contain features that tap or respond to our propensity to grant loyalty to tribes or reasonable facsimiles. The persistence of ethnic sentiments in a large-scale modern world that would seem to make them obsolete is an example (Glazer & Moynihan 1975). The ancient social instincts also retain important functions in the modern world. Families and personal friendships are important in every human social system.

The work-around hypothesis asserts that social instincts are part building-blocks and part constraints on the evolution of complex social systems. To evolve large-scale, complex social systems, culturally evolved strategies take advantage of whatever support the instincts offer, while coping as well as possible with the difficulty of raw material evolved for life in quite different sorts of societies. Families willingly take on the essential roles of biological reproduction and primary socialization. Appropriate larger-scale institutions can acceptably regulate their tendency to narrow loyalties and nepotistic subversion of group-favouring rules. Tribal-scale loyalties put deep emotion behind group enterprises, though the small scale of 'natural' tribes requires careful management if larger-scale objectives are not to be sacrificed. Large national and international (e.g. great religions) institutions develop ideologies of symbolically marked inclusion that often fairly successfully engage the tribal instincts on a much larger than tribal scale. The existence of contemporary societies handicapped by few loyalties outside the family (Banfield 1958) or by excessively powerful loyalties to small tribes (West 1941) remind us that work-arounds are awkward compromises that are difficult to achieve and easy to lose.

The most important cultural innovations required to support complex societies are command and control institutions that can systematically organize co-operation, co-ordination, and a division of labour in societies consisting of hundreds of thousands to hundreds of millions of people. Command and control institutions lead to more productive economies, more internal security, and better resistance to external aggression. Note that command and control are separable concepts. Command may aim at quite limited control. For example, a predatory conquest state may use command almost exclusively for the extraction of portable wealth, not for pro-social projects. Institutions often exert control without personal commands. Markets most famously control behaviour by price signals from a diffuse world of anonymous buyers and sellers. Market enthusiasts sometimes forget that command systems are generally needed to make markets function, ranging from mandatory use of calibrated weights and measures to central banks (Dahrendorf 1968: Chapter 8). The main types of work-arounds seem to us to be those detailed in the following:

Coercive dominance

The cynics' favourite mechanism for creating complex societies is command backed up by force. The conflict model of state formation has this character (Carneiro 1970). A society successful in war upon a neighbouring group can impose itself as a ruling class on the defeated if the defeated cannot flee, as farmers often cannot.

Elements of coercive dominance are no doubt necessary to make complex societies a going concern. Tribally legitimated self-help violence is a limited and expensive means of pro-social coercion. Complex human societies have to supplement the moralistic solidarity of tribal societies with formal police institutions. Otherwise, the large-scale benefits of co-operation, co-ordination, and division of labour would cease to exist in the face of selfish temptations to expropriate them by individuals, nepotists, cabals of reciprocators, organized predatory bands, and classes or castes with special access to means of coercion. At the same time, the need for organized coercion as an ultimate sanction creates roles, classes, and subcultures with the power to turn coercion to narrow advantage. Social institutions of some sort must police the police so that they will act in the larger interest to a measurable degree. Such policing is never perfect and, in the worst cases, can be very poor. The fact that leadership in complex systems always involves at least some economic inequality shows that narrow interests, rooted in individual selfishness, kinship, and, often, the tribal solidarity of the elite, always exert an influence. The use of coercion in complex societies offers excellent examples of the imperfections in social arrangements traceable to the ultimately irresolvable tension between selfish and pro-social instincts.

While coercive, exploitative elites are common enough, there are two reasons to suspect that no complex society can be based purely on coercion. The first problem is that coercion of any great mass of subordinates requires that the elite class or caste be itself a complex, co-operative venture. The second problem with pure coercion is that defeated and exploited peoples seldom accept subjugation as a permanent state of affairs without costly protest. Deep feelings of injustice generated by manifestly inequitable social arrangements move people to desperate acts, driving the cost of dominance to levels that cripple societies in the short run and often cannot be sustained in the long run (Kennedy 1987). Insko *et al.*'s (1983) experimental evolutionary analysis of coercive versus more pro-social leadership in laboratory micro-societies illustrates the degree to which dominated groups will chafe and rebel at their oppression. Durable conquests, such as those leading to the modern European national states, Han China, or the Roman Empire, leaven raw coercion with more pro-social institutions. The Confucian system in China and the Roman legal system in the West were far more sophisticated and durable institutions than the highly coercive systems sometimes set up by predatory conquerors and even domestic elites.

Segmentary hierarchy

Late Pleistocene societies were undoubtedly segmentary in the sense that supraband ethno-linguistic units served social functions, although they presumably lacked much formal political organization. The segmentary principle can serve the need for more command and control by hardening up lines of authority without disrupting the face-to-face nature of proximal leadership present in egalitarian societies. The Polynesian ranked lineage system illustrates how making political offices formally hereditary according to a kinship formula can help deepen and strengthen a command and control hierarchy (Kirch 1984; Sahlins 1963). A common method of deepening and strengthening the hierarchy of command and control in complex societies is to construct a formal nested hierarchy of offices, using various mixtures of ascription and achievement principles to staff the offices. Each level of the hierarchy replicates the structure of a hunting and gathering band. A leader at any level interacts mainly with a few near-equals at the next level down in the system. New leaders are usually recruited from the ranks of sub-leaders, often tapping informal leaders at that level. As Eibl-Eibesfeldt (1989: 314) remarks, even high-ranking leaders in modern hierarchies adopt much of the humble headman's deferential approach to leadership.

The hierarchical nesting of social units in complex societies gives rise to appreciable inefficiencies. In practice, brutal sergeants, incompetent colonels, vainglorious generals, and their ilk in other bureaucracies degrade the effectiveness of social organizations in complex societies. Squires (1986), elaborating on Tullock (1965), dissects the problems and potentials of modern hierarchical bureaucracies to perform consistently with leaders' intentions. Leaders in complex societies must convey orders downward, not just seek consensus among their comrades. Only very careful attention to detail can make subordinates responsive to the hierarchy's leaders without destroying their sense that these same leaders would have arisen by natural consensus without imposition from above. The chain of command is necessarily long in large complex societies, and remote leaders will not normally be able to exercise personal charisma over a mass of subordinates deeper down the hierarchy. Devolving substantial leadership responsibility to sub-leaders far down the chain of command is necessary to create small-scale leaders with face-to-face legitimacy. However, it potentially generates great friction if lower-level leaders either come to have different objectives from these of the upper leadership or are seen by followers as equally helpless pawns of remote leaders. Stratification often creates rigid boundaries so that natural leaders are denied promotion above a certain level, resulting in inefficient use of human resources and a fertile source of resentment to fuel social discontent.

Exploitation of symbolic systems

The high population density, division of labour, and improved communication made possible by the innovations of complex societies increased the scope for elaborating symbolic systems. The development of monumental architecture to serve mass ritual performances is one of the oldest archaeological markers of emerging complexity. Usually an established church or less formal ideological umbrella supports a complex society's institutions. At the same time, complex societies extensively exploit the symbolic ingroup instinct to delimit a diverse array of culturally defined subgroups, within which a good deal of co-operation is routinely achieved. Military organizations generally mark a set of middle-level, tribal-scale units with conspicuous badges of membership. A squad or platoon's solidarity can rest on bonds of reciprocity reinforced by pro-social leadership, but ship's companies, regiments, and divisions are made real by symbolic marking. Ethnic group-like sentiments in military organizations are often most strongly reinforced at the level of 1,000–10,000 or so men (British and German regiments, US divisions) (Kellett 1982: 112–17). Typical civilian symbolically marked units include regions (e.g. Swiss cantons), organized tribal elements (Garthwaite 1993), ethnic diasporas (Curtin 1984), castes (Gadgil & Guha 1992; Srinivas 1962), large economic enterprises (Fukuyama 1995), and civic organizations (Putnam 1993).

Many problems and conflicts revolve around symbolically marked groups in complex societies. Official dogmas often stultify desirable innovations and lead to bitter conflicts with heretics. Marked subgroups often have enough tribal cohesion to organize at the expense of the larger social system, as when lower-level military units arrange informal truces with the enemy or when ideologies of elite superiority support excessively exploitative institutions. A major difficulty with loyalties induced by appeals to shared symbolic culture is the very language-like productivity possible with this system. Language itself is a classic badge of an ethnic group. Dialect markers of social subgroups emerge rapidly along social fault-lines (Labov 1972). Charismatic innovators regularly launch new belief and prestige systems, which sometimes make radical claims on the allegiance of new members, sometimes make large claims at the expense of existing institutions, and sometimes grow explosively. Or, contrariwise, larger loyalties can arise, as in the case of modern nationalisms overriding smaller-scale loyalties, sometimes for better, sometimes for worse. The ongoing evolution of social systems can evolve in unpredictable, maladaptive directions by such processes. Gibbon (1776–88) attributed the decline and fall of Rome in part to the rise of Christianity (a timid and pacifistic ideology unsuited to empire, according to his notorious hypothesis). The worldwide growth of fundamentalist sects that challenge the institutions of modern states is a contemporary example (Marty & Appleby 1991; Roof & McKinney 1987). The contemporary Chinese state fears *Falun Gong*, whether with reason or not is hard to say. Resurgent ethnic loyalties recently wrecked Yugoslavia. The rise of the various 'isms' in some of the most powerful nation-states of the world made parts of the twentieth century a sanguinary hell. Ongoing cultural evolution is impossible to control, or at any rate impossible to control completely.

Legitimate institutions

At their most functional, symbolic institutions, together with effective leadership and smooth articulation of social segments, create a sense of living under a regime of tolerably fair laws and customs. Rationally administered bureaucracies, lively markets, the protection of socially beneficial property rights, and widespread participation in public affairs, provide public and private goods efficiently, along with a measure of protection of individual liberties. Individuals in modern societies typically feel themselves part of culturally labelled tribal-scale groups, such as local political party organizations, that have influence on the remotest leaders. In older complex societies, village councils, local notables, tribal chieftains, or religious leaders often hold courts open to humble petitioners. These local leaders in turn represent their communities to higher authorities. As long as most individuals feel that existing institutions are reasonably legitimate and that any felt needs for reform are achievable by means of ordinary political activities, there is considerable scope for collective social action.

On the other hand, individuals who do not trust the current institutional order's justness are liable to band together in revolutionary organizations, such as the terrorist groups of the contemporary world. Trust varies considerably in complex societies, and variation in trust is the main cause of differences in happiness across societies (Inglehart & Rabier 1986). Even the most efficient legitimate institutions known are prey to manipulation by small-scale organizations and cabals, the so-called special interests of modern democracies.

A test

Elsewhere we have used the differential performance of armies in the Second World War as a specific test of the work-around hypothesis (Richerson and Boyd 1999). We chose military organizations as a test because the extreme demands for personal sacrifice expected of modern soldiers in wartime exaggerate the conflict between individual interest (and loyalty to kin) and altruistic motivations to act on behalf of the larger society. In summary, the more successful institutions of complex societies go quite a way towards simulating the social institutions of simple societies. The German army in the Second World War outperformed Allied armies (on a man-for-man basis, controlling for the advantage of the Soviet army in numbers and the Western Allies' superior supply system and control of the air): 100 Germans could accomplish the same tasks that would require about 120 British or American troops or 200 Russians. This superiority, military analysts believe, came from the Germans' meticulous concern for the social-psychological needs of soldiers. German divisions were recruited on a territorial basis so that recruits shared their dialect and other symbols of regional identity. Thus care was taken to furnish soldiers with a tribal identity via identification with their regiments and divisions. Interestingly, although at least some soldiers were motivated by Hitler's bent ideological mission to 'save' Europe from the Jews and Bolsheviks, the army put little trust in the sustaining power of such sentiments and took care to create strong loyalties to divisions and regiments on a truly tribal scale. German training emphasized building solidarity with comrades, so personal bonds of loyalty were exploited. Unlike in the US army, those who trained together fought together. German face-to-face leaders were expected to minimize social distance from their men and to look out for their welfare at every turn. They gave orders in the form of objectives to be met, leaving the means of meeting them up to individuals. Physical coercion by leaders was common only in the Soviet army. The German army placed its best leaders in the front lines, whereas the American managerial approach demanded much talent in rear areas to organize the flow of supplies. German officers and NCOs were thus more like the informal natural leaders of hunting and gathering societies than was the case in Allied armies. In several relatively small but symbolic ways the German army expressed its concern for the welfare of individuals. Medals were awarded promptly for real combat accomplishments, an efficient field postal system kept soldiers in touch with their families, and hardship leaves were frequently granted, for example to help family if they had been bombed. By making individual soldiers feel well cared for as individuals and as participants in a tribal enterprise, the German army sustained their morale and exemplary performance even under the horrifying conditions of the Eastern Front.

THE ORIGINS OF AGRICULTURE EXPERIMENT

Several independent trajectories of subsistence intensification, often leading to agriculture, began during the Holocene (Richerson *et al*. in press). By intensification we mean a cycle of innovations in subsistence efficiency per unit of land leading to population growth that in turn leads to denser settlement per unit area of land. No plant-rich intensifications are known from the Pleistocene. Subsistence in the Pleistocene seems to have depended substantially on relatively high-quality animal and plant resources that held human populations to modest densities. Recent data from ice core climate proxies show that the last glacial climates were extremely hostile to agriculture — dry, low in atmospheric CO_2, and extremely variable on quite short timescales (Bradley 1999; Broecker 1995). We believe that these data suggest that agriculture was impossible under last-glacial conditions. Human populations appear to have been biologically quite modern in behaviour in most respects from Upper Paleolithic times forward (40,000–50,000 BP; Klein 1999). Population growth is a rapid process on timescales shorter than a millennium. Cultural evolution is a rapid process on timescales of ten millennia. If agriculture had been possible in the Pleistocene, it should have appeared before the Pleistocene–Holocene transition. The quite abrupt final amelioration of the climate at the onset of the Holocene 11,500 BP was followed immediately by the beginnings of plant-intensive resource use strategies in some areas, although the turn to plants was much later elsewhere. Almost all trajectories of subsistence intensification in the Holocene are progressive and eventually agriculture became the dominant strategy in all but the most marginal environments. The Polynesian expansion of the past 1,500 years and the European expansion of the past 500 years pioneered agriculture in the Pacific Islands, Australia, and large parts of western North America, the last substantial areas of the earth's surface favourable to it.

Two distinctive regimes for institutional evolution

Thus, evolution of human subsistence systems during the career of anatomically modern humans seems to divide quite neatly into two regimes: a Pleistocene regime of hunting and gathering subsistence and low population density, and a Holocene regime of increasingly agricultural subsistence and relatively high and rising population densities.

The dispersed resources and low mean density of populations in the Pleistocene meant that relatively few people could be aggregated together at any one time and place. The lack of domestic livestock meant that movement of goods on land would be limited to what humans could carry. No evidence of extensive use of boats to transport goods appears in the archaeological record of the Late Pleistocene, although some significant water crossings were necessary for people to reach Australia. Low-density, logistically limited human populations have small (but far from negligible, as we saw above) scope for exploiting returns to scale in co-operation, co-ordination, and division of labour and their institutions remain comparatively simple.

Intensified subsistence and higher population densities multiply the number of people and volume of commodities that societies can mobilize for economic and political purposes. Expanded exchange allows societies to exploit an expanded division of labour. Larger armies are possible to deal with external threats or to coerce neighbours. Expanding the number of people sharing a common language and customs will accelerate the spread of useful ideas. *Given appropriate institutions*, the denser societies made possible by agriculture can realize considerable returns to better exploitation of the potential of co-operation, co-ordination, and the division of labour. Corning (1983) elaborates the advantage of large-scale and greater complexity of social organization along these lines under his synergism hypothesis. Thus, in the Holocene, the origins of agriculture and its rising productivity over succeeding millennia at least permit the evolution of more complex societies.

A competitive ratchet

Intra- and inter-society competition put a sharp point on the potential for more complex societies. Holding the sophistication of institutions constant, marginal increases in subsistence productivity per unit of land will lead to denser or richer populations that can out-compete societies with less-intensive subsistence systems. Holding subsistence productivity constant, societies with marginally more sophisticated social organization will also out-compete rivals. Within groups, contending political interests with innovations that promise greater rewards for altered social organization can use either selfish or patriotic appeals to advance their cause. Successful reformers may entrench themselves in power for a considerable period. Malthusian growth will tend to convert increases in subsistence efficiency and security against depredations to greater population density, making losses of more complex institutions painful and further advance rewarding. Richerson *et al.* (in press) show that the rate limiting process for intensification trajectories must almost always be the rate of innovation of subsistence technology or subsistence-related social organization. At the observed rates of innovation, observed rates of population growth will always be rapid enough to sustain a high level of population pressure favouring further subsistence and social-organization innovations. Competition may be economic, political/military, or for the hearts and minds of people. Typically all three forms will operate simultaneously. In the Holocene, agriculture and complex social organization are, in the long run, compulsory. Thus, from the sixteenth to the nineteenth century, European populations settled many parts of the world and overwhelmed native populations with less efficient subsistence and less complex social organization. In regions such as Asia, where disparities of subsistence and social organization with the West were less striking, societies such as those of China, Japan, and India retained or reclaimed their political independence at the cost of humiliating exposure to western power and of borrowing many technical and social-organizational techniques from the West.

The tendencies for population to grow rapidly and for knowledge of advanced techniques to be retained somewhere act as pawls on the competitive ratchet. Even during the European Dark Ages, when the pawls slipped several cogs on the ratchet, the slide backward was halted and eventually reversed in a few hundred years.

Replications of the experiment

Agricultural subsistence evolved independently at least seven times in the Holocene and many more societies have acquired at least some key agricultural innovations by subsistence (Richerson *et al.* in press). Although none of these origins is earlier than the Early Holocene, many are much later. The trajectory of institutional evolution is similar. To take one benchmark, the origin of the state level of political organization began in Mesopotamia around 5,500 BP, but most regions are later, some much later (Feinman & Marcus 1998; Service 1975). For example, the Polynesian polities of Hawaii and Tonga-Samoa became complex chiefdoms on the cusp of the transition to states just before European contact (Kirch 1984). Pristine states evolved independently, perhaps ten or so times, in several parts of the world and traditions of statecraft in various regions evolved in substantial isolation for significant periods.

If our basic hypothesis is correct, the climatic shift at the Pleistocene–Holocene transition removed a tight constraint on the evolution of human subsistence systems and hence on institutional evolution. On the evidence of the competitive success of modern industrial societies, subsistence evolution has yet to exhaust the potential for more efficient subsistence inherent in agricultural production, and ongoing increases in the complexity of social institutions suggest that institutional evolution is still discovering more synergistic potential in human co-operation, co-ordination, and division of labour. The out-of-equilibrium progressive trend in human evolution over the past eleven millennia means that we can achieve a certain conceptual and probably empirical simplification of the problem of the evolution of institutions in the Holocene. We can assume a strong, worldwide tendency, driven by the competitive ratchet, towards societies at least as complex as current industrial societies. We can assume that changes in climate and similar non-social environmental factors play a small role in the Holocene. Granted these assumptions, we are left with three questions about subsistence and institutional evolution. First, why are rates of change so rapid in some areas (western Eurasia) and slow in others (western North America)? The competitive ratchet seems to have been routinely cranked faster in some places than others. What are the factors that limit the rate of cultural evolution in some cases relative to others? We shall argue that several processes can retard the rate of cultural evolution sufficiently to account for the observed rates of change. The second question is: how do we explain the multi-linear pattern of the evolution of institutional complexity? Although an upward trend of complexity characterizes most Holocene cultural traditions, the details of the trajectory vary considerably from case to case. The operation of the ratchet is very far from pulling all evolving social systems through the same stages; only relatively loose parallels exist between the cases. And third: Why does the ratchet sometimes slip some cogs? In no particular cultural tradition is progress even and steady. Episodes of temporary stagnation and regression are commonplace.

WHAT REGULATES THE TEMPO AND MODE OF INSTITUTIONAL EVOLUTION?

The overall pattern of subsistence intensification and increase in social complexity is clearly consistent with the hypothesis that agriculture and hence complex social institutions were impossible in the Pleistocene but eventually mandatory in the Holocene. The real test, however, is whether or not we can give a satisfactory account of the variation in the rate and sequence of cultural evolution. Work on this project is in its infancy, and what follows is only a brief sketch of the issues involved.

Geography may play a big role

Diamond (1997) argues that Eurasia has had the fastest rates of cultural evolution in the Holocene because of its size and to a lesser extent its orientation. Plausibly, the number of innovations that occur in a population increases with total population size and the flow of ideas between sub-populations. Since we know that the original centres of cultural innovation were relatively small compared with the areas to which they later spread, most societies acquired most complex cultural forms by diffusion. Societies isolated by geography will have few opportunities to acquire innovations from other societies. Contact of isolated areas with the larger world can have big impacts. The most isolated agricultural region in the world, Highland New Guinea, underwent an economic and social revolution in the past few centuries with the advent of American sweet potatoes, a crop that thrives in the cooler highlands above the malaria belt of lowland New Guinea (Wiessner and Tumu 1998). The Americas, though quite respectable in size, are oriented with their major axis north–south. Consequently innovations have mainly spread across lines of latitude from the homeland environment to quite different ones, unlike Eurasia where huge east–west expanses exist in each latitude belt. The pace of institutional change in Eurasian societies mirrors this region's early development of agriculture and the more rapid rate of subsistence intensification.

Climate change may play a small role

The Holocene climate is only invariant relative to the high-frequency, high-amplitude oscillations of the last glacial (Lamb 1977). For example, seasonality (difference between summer and winter insolation) was at a maximum near the beginning of the Holocene and has fallen since. The so-called 'Climatic Optimum', a broad period of warmer temperatures during the middle Holocene, caused a wetter Sahara, and the expansion of early pastoralism into what is now forbidding desert. The late medieval onset of the Little Ice Age caused the extinction of the Greenland Norse colony (Kleivan 1984). Agriculture at marginal altitudes in such places as the Andes seems to respond to Holocene climatic fluctuation (Kent 1987). The fluctuating success of state-level political systems in the cool, arid Lake Titicaca region is plausibly caused by wetter episodes permitting economies that support states, while these collapse or fade during arid periods. While the effect of Holocene climate fluctuations on regional sequences must always be kept in mind, the dominance of the underlying monotonic tendency to increase subsistence intensification and evolve more complex institutions seems likely to be driven by other processes.

Coevolutionary processes probably play a big role

The full exploitation of a revolutionary new subsistence system such as agriculture requires the evolution of domesticated strains of plants and animals. Human social institutions must undergo a revolution to cope with the increased population densities that follow from agricultural production. Human biology changes to cope with the novel dietary requirements of agricultural subsistence.

Agriculture requires pre-adapted plants and animals

In each centre of domestication, people domesticated only a handful of the wild plants that they formerly collected, and of this handful even fewer are widely adopted outside those centres. The same is true for domesticated livestock. Zohary and Hopf (1993) have listed some of the desirable features in plant domesticates. California has so many climatic, topographic, and ecological parallels with the precocious Fertile Crescent that its very tardy development of plant-intensive subsistence systems is a considerable puzzle. Diamond (1997), drawing on the work of Blumler (1992), notes that the Near Eastern region has a flora that is unusually rich in large-seeded grasses. California, by contrast, lacked large-seeded grasses, having not a single species that passed Blumler's criterion. Aside from obvious things such as large seed size, most Near Eastern domesticates had high rates of self-fertilization. This means that farmers can select desirable varieties and propagate them with little danger of gene flow from other varieties or from weedy relatives. Maize, by contrast, outcrosses at high rates. Perhaps the later and slower evolution of maize compared with Near Eastern domesticates is due to the difficulty of generating responses to selection in the face of gene flow from unselected populations (Diamond 1997: 137). Smith (1995) discusses the many constraints on potential animal domesticates.

Even in the most favourable cases, the evolution of new domesticates is not an instantaneous process. Blumler and Byrne (1991) identify the rate of evolution of domesticated characters such as non-dehiscence as one of the major unsolved problems of archaeobotany. Coevolution theorists such as Rindos (1984) imagine a long drawn-out period of modification leading up to the first cultivation, whereas Blumler and Byrne conclude that the rate of evolution of domesticates *may* be rapid, while stressing the uncertainties deriving from our poor understanding of the genetics and population genetics of domestication. Hillman and Davies' (1990) simulations indicate that the evolution of a tough rachis (the primary archaeological criterion of domestication) in inbreeding plants such as the wheats and barley could easily be so rapid as to be archaeologically invisible, as, indeed, it so far is. Their calculations also suggest that outcrossed plants, such as maize, will respond to cultivator selection pressures on the much longer timescales that Rindos and Diamond envision.

Humans have to adapt biologically to agricultural environments

While the transition from hunting and gathering to agriculture resulted in no genetic revolution in humans, a number of modest new biological adaptations were likely involved in becoming farmers. The best-documented case is the evolution of adult lactose absorption in human populations with long histories of dairying (Durham 1991). To some extent the relatively slow rate of human biological adaptation may act as a drag on the rate of cultural innovations leading to subsistence intensification and institutional advances.

Diseases limit population expansions, protect inter-regional diversity

McNeill (1976) and Crosby (1986) draw our attention to the coevolution of people and diseases. The increases in population density that resulted from the intensification of subsistence invited the evolution of epidemic diseases that could not spread at lower population densities. One result of this process is possibly to slow population growth to limits imposed by the evolution of cultural or genetic adaptations to diseases. For example, a suite of haemoglobins have arisen in different parts of the world that confer partial protection against malarial parasitism and these adaptations may have arisen only with the increases in human population densities associated with agriculture (Cavalli-Sforza *et al.* 1994). Cavalli-Sforza *et al.* estimate that it would take about 2,000 years for a new mutant haemoglobin variant to reach equilibrium in a population of 50,000 or so individuals (see also Gifford-Gonzales in press). Serious epidemics also have direct impacts upon social institutions when they carry away large numbers of occupants of crucial roles at the height of their powers. In such epidemics significant losses of institutional expertise could occur, directly setting back progressive evolution. Regional suites of diseases handicap immigrants and travellers, thus tending to isolate societies from the full effects of cultural diffusion.

Cultural evolutionary processes play a decisive role

The processes of cultural evolution may generally be more rapid than biological evolution, but cultural change often takes appreciable time. We (Boyd & Richerson 1985) view cultural evolution as a Darwinian process of descent with modification. Evidence about characteristic rates of modification is important for understanding the relative importance of various processes in cultural evolution. In one limit, the conservative, blind, transmission of cultural variants from parents to offspring, the main adaptive force on cultural variants would be natural selection, and rates of cultural evolution would approximate those of genes. At the other extreme, humans may pick and chose among any of the cultural variants available in the community and may use cognitive strategies to generate novel behaviours directly in light of environmental contingencies (Borgerhoff Mulder *et al.* 1997). In the limit of economists' omniscient rational actors, evolutionary adjustments are modelled as if they are instantaneous. We believe that for many cultural traits human decisions have relatively weak effects in the short run and at the individual level, although they can be powerful when integrated over many people and appreciable spans of time. Archaeological and historical data on the rates of change in different domains of culture will be some of the most important evidence to muster to understand the tempo and mode of cultural evolution. Much work remains to be done before we understand the regulation of rates of cultural evolution, but some preliminary speculation is possible.

New technological complexes evolve with difficulty

One problem that will tend to slow the rate of cultural (and organic) evolution is the sheer complexity of adaptive design problems. As engineers have discovered when studying the design of complex functional systems, discovering optimal designs is quite difficult. Blind search algorithms often get stuck on local optima, of which complex design problems often have very many. Piecemeal improvements at the margin are not guaranteed to find globally optimal adaptations by myopic search. Yet, myopic searches are what Darwinian processes do (Boyd & Richerson 1992b). Even modern engineering approaches to design, for all their sophistication, are more limited by myopic cut and try than engineers would like.

Parallel problems are probably rife in human subsistence systems. The shift to plant-rich diets is complicated because plant foods are typically deficient in essential amino acids and vitamins, have toxic compounds to protect them from herbivore attack, and are labour-intensive to prepare. Finding a mix of plant and animal foods that provides adequate diet at a feasible labour cost is not a trivial problem. For example, New World farmers eventually discovered that boiling maize in wood ashes improved its nutritional value. The hot alkaline solution breaks down an otherwise indigestible seed coat protein that contains some lysine, an amino acid that is low in maize relative to human requirements (Katz *et al.* 1974). Hominy and *masa harina*, the corn flour used to make tortillas, are forms of alkali-treated maize. The value of this practice could not have been obvious to its inventors or later adopters, yet most American populations that made heavy use of maize employed it. The dates of origin and spread of alkali cooking are not known. It has not been reinvented in Africa even though many African populations have used maize as a staple for centuries.

New social institutions evolve with difficulty

An excellent case can be made that the rate of institutional innovation is more often limiting than the rate of innovation of technology. As anthropologists and sociologists such as Julian Steward (1955) have long emphasized, human economies are social economies. Even in the simplest human societies, hunting and gathering is never a solitary occupation. At the minimum, such societies have division of labour between men and women. Hunting is typically a co-operative venture. The unpredictable nature of hunting returns typically favours risk-sharing at the level of bands composed of a few co-operating families because most hunters are successful only every week or so (Winterhalder 1986). Portions of kills are distributed widely, sometimes exactly equally, among band members.

The deployment of new technology requires changes in social institutions to make best use of innovations, often at the expense of entrenched interests, as Marx argued. The increasing scale of social institutions associated with rising population densities during the Holocene has dramatically reshaped human social life. Richerson and Boyd (1998, 1999) discuss the complex problems involved in evolutionary trajectory from small-scale, egalitarian societies to large-scale complex societies with stratification and hierarchical political systems. For example, even the first steps of intensification required significant social changes. Gathering is generally the province of women and hunting of men. Male prestige systems are often based on hunting success. A shift to plant resources requires scheduling activities around women's work rather than men's pursuit of prestige. Using more plants will conflict with men's preferences as driven by a desire for hunting success; it will require a certain degree of women's liberation to intensify subsistence. Since men generally dominate women in group decision-making ('egalitarian' small-scale societies seldom grant women equal political rights), male chauvinism will tend to limit intensification. Bettinger and Baumhoff (1982) argue that the spread of Numic speakers across the Great Basin a few hundred years ago was the result of the development of a plant-intensive subsistence system in the Owens Valley. Apparently, the groups that specialized in the hunt would not or could not shift to the more productive economy to defend themselves, perhaps because males clung to the outmoded, plant-poor, subsistence. Winterhalder and Goland (1997) use optimal foraging analysis to argue that the shift from foraging to agriculture would have required a substantial shift in risk-management institutions, from minimizing risk by intraband and interband sharing to reducing risk by field dispersal by individual families. Some ethnographically known Eastern Woodland societies that mixed farming and hunting, for example the Huron, seem not to have made this transition and to have suffered frequent catastrophic food shortages.

Institutional evolution no doubt involves complex design problems. For example, Blanton (1998) describes some of the alternative sources of power in archaic states. He notes that archaic states differ widely in time and space as their evolution wanders about in a large space of alternative social institutions. Thus, the Classical Greek system of small egalitarian city-states with participation in governance by male citizens was a far different system from others, such as that in Egypt, with divine royal leaders almost from the time of its inception as a state or the bureaucracies that were common in western Asia. Philip, Alexander, and their successors substantially rebuilt the Greek state along western Asian lines in order to conquer and administer empires in Asia.

Much of the medium-term change in archaic and classical state institutions seems to involve wandering about in a large design space without discovering any decisively superior new institutional arrangements (Feinman 1998; Marcus 1998).

The spread of complex social institutions by diffusion is arguably more difficult than the diffusion of technological innovations. Social institutions violate four of the conditions that tend to facilitate diffusion (Rogers 1983). Foreign social institutions are often (1) not compatible with existing institutions, (2) complex, (3) difficult to observe, and (4) difficult to try out on a small scale.

Thus the evolution of social institutions rather than technology will tend to be the rate-limiting step of the intensification process. For example, North and Thomas (1973) argue that new and better systems of property rights set off the modern industrial revolution rather than the easier task of technical invention itself. A major revolution in property rights is also likely to be necessary for intensive hunting and gathering and agriculture to occur (Bettinger 1999). Slow diffusion also means that historical differences in social organization can be quite persistent, even though one form of organization is inferior. As a result, the comparative history of the social institutions of intensifying societies exhibits many examples of societies gaining a persistent competitive advantage over others in one dimension or another because they possess an institutional innovation that their competitors do not acquire. For example, the Chinese merit-based bureaucratic system of government was established at the expense of the landed aristocracy, beginning in the Han dynasty (2,200 BP) and completed in the Tang (1,400 BP) (Fairbank 1992). This system has become widespread only in the modern era and is still quite imperfectly operated in many societies.

To the extent that games of co-ordination are important in social organization, changes from one co-ordination solution to another may be greatly inhibited. Games of co-ordination are those, such as which side of the road to drive on, where it matters a great deal that everyone agrees on a single solution but less on which solution is chosen (Sugden 1986). Notoriously, armies with divided command are defeated. A poor general's plan formulated promptly and obeyed without question by all is usually superior to two good generals' plans needing long negotiations to reconcile or leaving subordinates with a choice of whose plan to follow. We care less about whether gold, silver, or paper money is legal tender than we care about having a single standard. Many if not most social institutions probably have strong elements of co-ordination. Take marriage rules. Some societies allow successful men to marry several wives while others forbid the practice. One system may or may not be intrinsically better, but everyone is better off playing by one set of rules. Since the strategies appropriate for one possibility are quite different from those appropriate for the other, marriage partners would like agreement on the ground rules of marriage up front to save costly negotiation or worse later on. Hence many institutions are in the form of a socially policed norm or standard contract ('love, honour, cherish and obey until death do us part') solving what seems as though it ought to be a private co-ordination problem. However, except in pure cases, different co-ordination equilibria will also have different average payoffs and different distributions of payoffs than others. Even if most agree that a society can profitably shift from one simple pure co-ordination equilibrium to another (as when the Swedes switched from driving on the left to the right a couple of decades ago to conform to their neighbours' practices) the change is not simple to orchestrate. A US university voted recently not to switch from the quarter to the semester system despite a widespread recognition that a mistake was made 30 years ago when the quarter system was instituted. Large, uncertain costs that many semester-friendly faculty reckoned would attend such a switch caused them to vote no. Larger-scale changes, such as the Russian transition from a soviet to a capitalist economy, face huge problems that are plausibly the result of the need to renegotiate solutions to a large number of games of co-ordination as much as of any other cause.

Design complexity, importance of co-ordination, slow evolution, limited diffusion, and difficulty of co-ordination shifts probably conspire to make the evolution of social institutions highly historically contingent. The multilinear pattern of evolution of social complexity could result from two causes. Societies could be evolving from diverse starting points towards a single common optimal state surrounded by a smooth 'topography' which optimizing evolutionary processes are climbing towards the summit. Or societies could be evolving up a complex topography with many local optima and many potential pathways towards higher peaks. In this case, even if societies start out at very similar initial points, they will tend to diverge with time. We believe that at least part of the historical contingency in cultural evolution is due to slow evolution on complex topographies (Boyd & Richerson 1992b).

Ideology may play a role

Non-utilitarian processes may strongly influence the evolution of fads, fashions, and belief systems. Such forces are susceptible to feedback and runaway dynamics that defy common sense (Boyd & Richerson 1985: Chapter 8). The links between belief systems and subsistence are nevertheless incontestably strong. To build a cathedral requires an economy that produces surpluses that can be devoted to grand gestures on the part of the faithful. The moral precepts inculcated by the clergy in the cathedral underpin the institutions that in turn regulate the economy. Arguably, ideological innovations often drive economic change. Recall Max Weber's classical argument about the role of Calvinism in the rise of capitalism.

Complex social systems are vulnerable

We suggest that the fragility of institutions derives from compromises and trade-offs that are caused by conflicts between the functional demands of large-scale organizations and the trajectory of small-scale cultural evolution often driven by psychological forces rooted in the ancient and tribal social instincts. The evolution of work-arounds seldom results in perfect adaptations. Resistance to the pull of the ratchet can increase sharply when external pressures, such as competition from other societies, demographic catastrophes, or internal processes, for example, the evolution of a new religion, put weak work-arounds in jeopardy. All complex societies may have weak work-arounds lurking among their institutions. As we noted above, each of the major types of institutional work-arounds has defects that lead to intra-societal conflict. Small-scale societies have appreciable crudities at least in part deriving from conflicts, both intra-psychic and political, between individual and kinship interests and the larger tribe (Edgerton 1992). If our argument is correct, larger-scale societies do not eliminate these conflicts but add to them manifold opportunities for conflict between different elements of the larger system. Even the best of such systems current at any one time are full of crudities and the worst are often highly dysfunctional. A considerable vulnerability to crisis, change without progress, setback, and collapse is inherent in an evolutionary system subject to strong evolutionary forces operating at different levels.

Maynard Smith and Szathmáry (1995) treat the rise of human ultra-sociality as analogous to other major evolutionary transformations in the history of life. As with our tribal social instincts and work-around proposals, the key feature of transitions from, say, cellular grade organisms to multicellular ones is the improbable and rare origin of a system in which group selection works at a larger scale to suppress conflict at the smaller and eventually to perfect the larger-scale super-organisms. Actually, 'perfect' is too strong a word: distinct traces of conflict remain in multicellular organisms and honeybee colonies too. We suggest that human societies are recently evolved and remain rather crude super-organisms, heavily burdened by conflict between lower- and higher-level functions and not infrequently undone by them. Outside the realm of utopian speculation and science fiction, there does not appear to be an easy solution. Muddle along is the rule, pulled on the trajectory towards more social complexity by the competitive ratchet.

Changes in the rate of cultural evolution and the sizes of cultural repertoires

Rates of social and technical evolution appear to be rising towards the present. Modern individuals know more than did their ancestors, and social complexity has increased. The cultural evolution in the Holocene began at a stately pace. Not for some 6,000 years after the initial domestication of plants and animals in south-western Asia did the first state-level societies decisively transcend the tribal-scale roots of human sociality. Tribes, city-states, and empires competed to govern Eurasia for another 4,500 years while the first states emerged in the New World and Africa. The rise of the West over the past millennium has brought revolutions in subsistence and social organization, particularly during the past half-millennium. Even in Eurasia, the last pastoral and hunting tribes of the interior were only defeated by Chinese and Russian armies with firearms a couple of centuries ago. Only in the past century or two has cultural evolution been sufficiently rapid that almost everyone is aware of major changes within their lifetime. Malthus, writing around the turn of the nineteenth century, still regarded technical innovation as quite slow, on quite sound empirical grounds. Only a couple of decades after his death would cautious empiricists have good grounds to argue that the industrial revolution was something new under the sun (Lindert 1985). The accelerating growth of the global population is a product of these changes and the curve of population growth is one reasonable overall index of cultural change. Another is the increasing division of labour. Innovations on the subsistence side at first rather gradually, and then later very rapidly, reduced the personnel devoted to agricultural production and shifted labour into an expanding list of mercantile, manufacturing, government, and service occupations.

The reasons for the accelerating rate of increase are likely to be several. First, the sheer increase in numbers of people must have some effect on the supply of innovations. Second, the invention of writing and mathematics provided tools for supplementing memories, for aiding the application of rationality, and for the long-distance communication of ideas. Scribes in small numbers first used their new skills to manage state supply depots, taxation, and land mensuration. Only gradually did procedures in different fields become written and mathematics come to be used to solve an expanding array of problems. Third, books ultimately became a means of both conserving and communicating ideas, at first only to an educated elite. Fourth, quite recently, the mass of people in many societies became literate and numerate, allowing most people to take up occupations dependent upon prolonged formal education, policy and procedure handbooks, technical manuals, reference books, and elaborate calculations. Fifth, the rise of cheap mass communication, beginning with the printed book, has given individuals access to ever-larger stores of information. The internet promises to give everyone able to operate a workstation access to all the public information in the world. Donald (1991) counts the spread of literacy and numeracy as a mental revolution on the same scale as the evolution of imitation and spoken language. Sixth, institutions dedicated to deliberately promoting technical and social change have grown much more sophisticated. Boehm (1996) argues that even acephalous societies usually have legitimate, customary institutions by which the society can reach a consensus on actions to take in emergencies, such as the threat of war or famine (see also Turner 1995: 16–17).

Institutions organized as a matter of social policy to further change continue to increase perceptibly in scope and sophistication. Institutions such as patents that give innovators a socially regulated property right in their inventions ushered in the industrial revolution. Private companies invest in new technology under the eye of government regulators from about the turn of the twentieth century. Government bureaucracies began to conduct useful research using funds from the public purse in a small way in the nineteenth century. Research universities recruit some of the best minds available, place them in an intellectual hothouse, and reward scholars for new ideas of whatever kind they are prepared to pursue. Masses of young people are educated by such innovators and their students, especially during the past 50 years. The military's general staffs, modelled on the impressively effective Prussian/ German general staff developed by Scharnhorst, Gneisenau and their students such as Clausewitz, drove military modernization at impressive rates in the nineteenth and twentieth centuries. Dupuy (1977) argues that the effectiveness of German soldiers in the Second World War was in largest part a product of the German general staff drawing more appropriate lessons from the First World War and implementing more thorough reforms than did competing general staffs. Development institutions such as agricultural extension services and teaching hospitals move innovations in some fields from the university to the farm or doctor's office at a smart pace. Think-tanks ponder public policy in the light of research, national academies of science craft white papers based on elaborate searches for expert consensus, and legislatures hold hearings trying to match the desires of constituents with the findings of the experts in order to produce new policies and programmes.

WE HAVE A SHADOWY OUTLINE OF THE TEMPO AND MODE OF THE EVOLUTION OF INSTITUTIONS

The large, rapid change in environment at the Pleistocene–Holocene transition set off the trend of subsistence intensification and institutional complexity of which modern industrial innovations are just the latest examples. If our hypothesis is correct, the reduction in climate variability and increase in CO_2 content of the atmosphere, together with increases in rainfall, rather abruptly changed the Earth from a regime where agriculture was everywhere impossible to one where it was possible in many places. Since groups that utilize efficient, plant-rich subsistence systems and deploy the resulting larger population more effectively will normally out-compete groups that make less efficient use of land and people, the Holocene has been characterized by a persistent tendency towards subsistence intensification and growth in institutional sophistication and complexity. The diversity of trajectories taken by the various regional human sub-populations since ≈11,600 BP are natural experiments that will help us elucidate the factors controlling the tempo and mode of cultural evolution leading to more efficient subsistence systems and the more complex societies these systems support. A long list of processes interacted to regulate the trajectory of subsistence intensification, population growth, and institutional change that the world's societies have followed in the Holocene. Social scientists are in the habit of treating these processes as mutually exclusive hypotheses. They seem to us to be competing, but certainly not mutually exclusive. Many are not routinely given any attention in the historical social sciences. At the level of qualitative empiricism, tossing any one out entirely leaves puzzles that are hard to account for and produces a caricature of the actual record of change. If this conclusion is correct, the task for historically minded social scientists is to refine estimates of the rates of change that are attributable to the various evolutionary processes and to estimate how those rates change as a function of natural and sociocultural circumstances. We lack a quantitative understanding of the burden of flawed work-arounds and other features of complexity that retard and locally reverse tendencies to greater complexity. We only incompletely understand the processes generating historical contingency.

The present very high rates of technical and institutional evolution are a problem of immense applied importance. While some observers are complacent about current trends (Fukuyama 1992), others worry. For example, our headlong quest for increased material prosperity that guides so much current calculated institutional change not only takes great risks of environmental deterioration and a hard landing on the path to sustainability, but seems flawed from the point of view of satisfying human needs and wants (Easterlin 1995; Frank and Cook 1995).

THE END OF COMPLEX SOCIETIES?

Those who are familiar with the Pleistocene often remark that the Holocene is just the 'present interglacial'. The return of climate variation on the scale that characterized the last glacial is quite likely if current ideas about the Milankovich driving forces of the Pleistocene are correct. Sustaining agriculture and complex societies under conditions of much higher amplitude and more frequent environmental variation than farmers currently cope with would be a very considerable technical challenge. At the very best, lower CO_2 concentrations and lower world average precipitation suggest that world average agricultural output would fall considerably.

In one sense, though, the Holocene is not just another interglacial. Petit *et al.* (1999) suggest that it may be uniquely long, although decidedly cooler than the maximum temperatures of the previous four interglacials, at least in continental Antarctica. Current anthropogenic global warming via greenhouse gases threatens to elevate world temperatures to levels that in past interglacials apparently triggered a large feedback effect producing a relatively rapid decline towards glacial conditions. The Arctic Ocean ice pack is currently thinning very rapidly (Kerr 1999). A dark, open Arctic Ocean would dramatically increase the heat income at high northern latitudes, and have large, unpredictable impacts on the Earth's climate system. No one can yet estimate the risks we are taking of a rapid return to a colder, drier, more variable environment with less CO_2, nor evaluate exactly the threat such conditions imply for the continuation of agricultural production. Nevertheless, the intrinsic instability of the Pleistocene climate system, and the degree to which agriculture is dependent upon the unusually long Holocene stable period, should give one pause for thought (Broecker 1997).

Of course, our sophisticated understanding of the natural world and our ability to turn that understanding into purposive collective action must not be underestimated. Human societies are perhaps indeed a major transition in evolution as Maynard Smith and Szathmáry (1995) argue, but to our way of thinking this transformation is still a work in progress. Given the manifest remaining crudities of our social systems, the power of our own destructive inventions, and the potential of glacial climates and other forces of nature to wreak havoc on human plans, completing the human transition promises to be a near-run thing.

Note. We thank Bryan Vila for a close reading of the manuscript and his many thoughtful comments, and Garry Runciman for his invitation to participate and for his comments on our chapter.

REFERENCES

ALEXANDER, J.C., GIESEN, B., MÜNCH, R., & SMELSER, N.J. 1987: *The Micro-Macro Link* (Los Angeles, University of California Press).

ALEXANDER, R.D. 1974: The evolution of social behaviour. *Annual Review of Ecology and Systematics* 5, 325–83.

ALEXANDER, R.D. 1987: *The Biology of Moral Systems* (New York, Aldine de Gruyter).

ARNOLD, J.E. 1996: The archaeology of complex hunter-gatherers. *Journal of Archaeological Method and Theory* 3, 77–126.

AXELROD, R. & HAMILTON, W.D. 1981: The evolution of co-operation. *Science* 211, 1390–6.

BANFIELD, E.C. 1958: *The Moral Basis of a Backward Society* (Glencoe, IL, Free Press).

BAUM, W.B. 1994: *Understanding Behaviourism: Science, Behaviour, and Culture* (New York Harper Collins).

BEAN, L.J. 1978: Social organization. In Heizer, R.F. (ed.), *Handbook of North American Indians. Volume 8: California* (Washington DC, Smithsonian Institution Press), 673–82.

BETTINGER, R.L. 1991: *Hunter-Gatherers: Archaeological and Evolutionary Theory* (New York Plenum).

BETTINGER, R.L. 1999: From traveler to processor: regional trajectories of hunter-gatherer sedentism in the Inyo-Mono Region, California. In Billman, B.R. & Feinman, G.M. (eds.), *Settlement Pattern Studies in the Americas: Fifty Years Since Viru* (Washington DC, Smithsonian Institution Press), 39–55.

BETTINGER, R.L. & BAUMHOFF, M.A. 1982: The Numic spread: Great Basin cultures in competition. *American Antiquity* 47, 485–503.

BLANTON, R.E. 1998: Beyond centralization: steps toward a theory of egalitarian behaviour in archaic states. In Feinman, G.M. & Marcus, J. (eds.), *Archaic States* (Santa Fe, NM, School of American Research Press),135–72.

BLUMLER, M.A. 1992: 'Seed Weight and Environment in Mediterranean-type Grasslands in California and Israel'. PhD thesis, University of California, Berkeley.

BLUMLER, M.A. & BYRNE, R. 1991: The ecological genetics of domestication and the origins of agriculture. *Current Anthropology* 32, 23–53.

BOEHM, C. 1993: Egalitarian society and reverse dominance hierarchy. *Current Anthropology* 34, 227–54.

BOEHM, C. 1996: Emergency decisions, cultural selection mechanics, and group selection. *Current Anthropology* 37, 763–93.

BORGERHOFF MULDER, M., RICHERSON, P.J., THORNHILL, N.W. & VOLAND, E. 1997: The place of behavioural ecological anthropology in evolutionary social science. In Weingart, P. Mitchell, S.D., Richerson, P.J. & Maasen, S. (eds.), *Human By Nature — Between the Social Sciences and Biology* (Hillsdale, NJ, Lawrence Erlbaum), 253–82.

BOYD, R. & RICHERSON, P.J. 1985: *Culture and the Evolutionary Process* (Chicago, University of Chicago Press).

BOYD, R. & RICHERSON, P.J. 1987: The evolution of ethnic markers. *Cultural Anthropology* 2, 65–79.

BOYD, R. & RICHERSON, P.J. 1992a: Punishment allows the evolution of co-operation (or anything else) in sizable groups. *Ethology and Sociobiology* 13,171–95.

BOYD, R, & RICHERSON, P.J. 1992b: How microevolutionary processes give rise to history. In Nitecki, M.H. & D.V. (eds.), *History and Evolution* (Albany, State University Press of New York), 179–209.

BOYD, R. & RICHERSON, P.J. 1993: Rationality, imitation, and tradition. In Day, R.H. & Chen, P. (eds.), *Nonlinear Dynamics and Evolutionary Economics* (New York, Oxford University Press). pp. 131–49.

BOYD, R. & RICHERSON, P.J. 1996: Why culture is common but cultural evolution is rare. *Proceedings of the British Academy* 88, 77–93.

BOYD, R. & RICHERSON, P.J. 2000: Memes: universal acid or better mousetrap? In Aunger R. (ed.), *Darwinizing Culture* (Cambridge, Cambridge University Press), 143–62.

BRADLEY, R.S. 1999: *Paleoclimatology: Reconstructing Climates of the Quaternary* (San Diego, Academic Press).

BROECKER, W.S. 1995: *The Glacial World According to Wally* (Lamont-Doherty Earth Observatory of Columbia University, Palisades, NY).

BROECKER, W.S. 1997: Thermohaline circulation, the achilles heel of our climate system: will man-made CO_2 upset the current balance? *Science* 178, 1582–8.

CAMPBELL, D.T. 1965: Variation and selective retention in sociocultural evolution. In Barringer, H.R. Blanksten, G.I. & Mack, R.W. (eds.), *Social Change in Developing Areas: a Reinterpretation of Evolutionary Theory* (Cambridge MA, Schenkman), 19–49.

CARNEIRO, R.L. 1970: A theory of the origin of the state. *Science* 169, 733–8.

CAVALLI-SFORZA, L.L. & FELDMAN, M.W. 1973: Models for cultural inheritance: I. Group mean and within group variation. *Theoretical Population Biology* 4, 42–55.

CAVALLI-SFORZA, L.L. & FELDMAN, M.W. 1981: *Cultural Transmission and Evolution: a Quantitative Approach* (Princeton, Princeton University Press).

CAVALLI-SFORZA, L.L., MENOZZI, P. & PIAZZA, A. 1994: *The History and Geography of Human Genes* (Princeton, Princeton University Press).

COLEMAN, J.S. 1990: *Foundations of Social Theory* (Cambridge, MA, Harvard University Press).

CORNING, P.A. 1983: *The Synergism Hypothesis: a Theory of Progressive Evolution* (New York, McGraw-Hill).

CROSBY, A.W. 1986: *Ecological Imperialism: the Biological Expansion of Europe, 900–1900* (Cambridge, Cambridge University Press).

CURTIN, P.D. 1984: *Cross-Cultural Trade in World History* (Cambridge, Cambridge University Press).

DAHRENDORF, R. 1968: *Essays in the Theory of Society* (Stanford, Stanford University Press).

DALY, M. & WILSON, M. 1988: *Homicide* (New York, Aldine de Gruyter).

DARWIN, C. 1874 [1902]: *The Descent of Man*, 2nd edn. (New York, American Home Library).

DIAMOND, J. 1997: *Guns, Germs, and Steel* (New York, Norton).

DONALD, M. 1991: *Origins of the Modern Mind: Three Stages in the Evolution of Culture and Cognition* (Cambridge, MA, Harvard University Press).

DUPUY, T.N. 1977: *A Genius for War: the German Army and General Staff, 1807–1945* (Englewood Cliffs, NJ, Prentice-Hall).

DURHAM, W.H. 1991: *Coevolution: Genes, Culture, and Human Diversity* (Stanford, Stanford University Press).

EASTERLIN, R.A. 1995: Will raising the incomes of all increase the happiness of all? *Journal of Economic Behaviour and Organization* 27, 35–47.

EDGERTON, R.B. 1992: *Sick Societies: Challenging the Myth of Primitive Harmony* (New York, Maxwell Macmillan).

EIBL-EIBESFELDT, I. 1982: Warfare, man's indoctrinability, and group selection. *Zeitschrift für Tierpsychologie* 67, 177–98.

EIBL-EIBESFELDT, I. 1989: *Human Ethology* (New York, Aldine de Gruyter).

FAIRBANK, J.K. 1992: *China: a New History* (Cambridge, MA, Harvard University Press).

FEINMAN, G.M. 1998: Scale and social organization: perspectives on the archaic state. In Feinman, G.M. & Marcus, J. (eds.), *Archaic States* (Santa Fe, NM, School of American Research Press). 95–133.

FEINMAN, G.M. & MARCUS, J. 1998: *Archaic States* (Santa Fe, NM, School of American Research Press).

FRANK, R.H. & COOK, P.J. 1995: *The Winner-Take-All Society* (New York, Free Press).

FREUD, S. 1930: *Civilization and Its Discontents* (New York, Norton).

FUKUYAMA, F. 1992: *The End of History and the Last Man* (New York, Free Press).

FUKUYAMA, F. 1995: *Trust: the Social Virtues and the Creation of Prosperity* (New York, Free Press).

GADGIL, M. & GUHA, R. 1992: *This Fissured Land: an Ecological History of India* (Delhi, Oxford University Press).

GARTHWAITE, G.R. 1993: Reimagined internal frontiers: tribes and nationalism — Bakhtiari and Kurds. In Eichelman, D.F. (ed.), *Russia's Muslim Frontiers: New Directions in Cross-Cultural Analysis* (Bloomington, Indiana University Press), 130–48.

GIBBON, E. 1776–88 [1952]: *The Decline and Fall of the Roman Empire* (London, Penguin).

GIFFORD-GONZALEZ, D. in press: Animal disease challenges to the emergence of pastoralism in sub-Saharan Africa. *African Archaelogical Review*.

GLAZER, N. and MOYNIHAN, D.P. 1975: *Ethnicity: Theory and Experience* (Cambridge, MA, Harvard University Press).

HAMILTON, W.D. 1964: The genetical evolution of social behaviour I, II. *Journal of Theorectical Biology* 7, 1–52.

HAMILTON, W.D. 1975: Innate social aptitudes of man: an approach from evolutionary genetics. In Fox, R. (ed.), *Biosocial Anthropology* (London, Malaby), 133–55.

HENRICH, J. & BOYD, R.1998: The evolution of conformist transmission and the emergence of between-group differences. *Evolution and Human Behaviour* 19, 215–41.

HILLMAN, G.C., & DAVIES, M.S. 1990: Measured domestication rates in wild wheats and barley under primitive cultivation, and their archaeological implications. *Journal of World Prehistory* 4, 157–222.

INGLEHART, R. & RABIER, J.-R. 1986: Aspirations adapt to situations — but why are the Belgians so much happier than the French? A cross-cultural analysis of the subjective quality of life. In Andrews, F.M. (ed.), *Research on the Quality of Life* (University of Michigan, Ann Arbor, Survey Research Center, Institute for Social Research), 1–56.

INSKO, C.A., GILMORE, R., DRENAN, S., LIPSITZ, A., MOEHLE, D. & THIBAUT, J. 1983: Trade versus expropriation in open groups: a comparison of two types of social power. *Journal of Personality and Social Psychology* 44, 977–99.

KATZ, S.H., HEDIGER, M.L. & VALLEROY, L.A. 1974: Traditional maize processing techniques in the New World. *Science* 184, 765–73.

KELLETT, A. 1982: *Combat Motivation: the Behaviour of Soldiers in Battle* (Boston, Kluwer Nijhoff).

KELLY, R.L. 1995: *The Foraging Spectrum: Diversity in Hunter-Gatherer Lifeways* (Washington DC, Smithsonian Institution Press).

KENNEDY, P. 1987: *The Rise and Fall of Great Powers: Economic Change and Military Conflict from 1500 to 2000* (New York, Random House).

KENT, J.D. 1987: Periodic aridity and prehispanic Titicaca Basin settlement patterns. In Browman, D.L. (ed.), *Arid Land Use Strategies and Risk Management in the Andes* (Boulder, CO, Westview), 297–314.

KERR, R.A. 1999: Will the Arctic Ocean lose all its ice? *Nature* 286, 128.

KIRCH, P.V. 1984: *The Evolution of Polynesian Chiefdoms* (Cambridge, Cambridge University Press).

KLEIN, R.G. 1999: *The Human Career: Human Biological and Cultural Origins,* 2nd edn (Chicago, University of Chicago Press).

KLEIVAN, I. 1984: History of Norse Greenland. In Damas, D. (ed.), *Handbook of North American Indians 5: Arctic* (Washington, Smithsonian Institution), 549–55.

LABOV, W. 1972: *Sociolinguistic Patterns* (Philadelphia, University of Pennsylvania Press).

LAMB, H.H. 1977: *Climatic History and the Future* (Princeton, Princeton University Press).

LINDERT, P.H. 1985: English population, wages, and prices: 1541–1913. *Journal of Interdisciplinary History* 15, 609–34.

LUMSDEN, C.J. & WILSON, E.O. 1981: *Genes, Mind, and Culture: the Coevolutionary Process* (Cambridge, MA, Harvard University Press).

MARCUS, J. 1998: The peaks and valleys of ancient states: an extension of the dynamic model. In Feinman, G.M. & Marcus, J. (eds.), *Archaic States* (Santa Fe, NM, School of American Research Press), 59–94.

McELREATH, R., BOYD, R. & RICHERSON, P.J. no date: 'Shared norms can lead to the evolution of ethnic markers'. Unpublished manuscript.

McNEILL, W.H. 1976: *Plagues and Peoples* (Anchor, Garden City).

MARTY, M.E. & APPLEBY, R.S. 1991: *Fundamentalisms Observed* (Chicago, University of Chicago Press).

MAYNARD SMITH, J. & SZATHMÁRY, E. 1995: *The Major Transitions in Evolution* (Oxford, Oxford University Press).

NORTH, D.C. & THOMAS, R.P. 1973: *The Rise of the Western World: a New Economic History* (Cambridge, Cambridge University Press).

PETIT, J.R., JOUZEL, J., RAYNAUD, D., BARKOV, N.I., BARNOLA, J-M., BASILE, J., BENDER, M., CAPPELLAZ, J., DAVIS, M., DELAYGUE, G., DELMOTTE, M., KOTLYAKOV, V.M., LEGRAND, M., LIPENKOV, V.Y., LORIUS, C., PÉPIN, L., RITZ, C., SALTZMAN, E. & STIEVENARD, M. 1999: Climate and atmospheric history of the past 420,000 years from the Vostok ice core, Antarctica. *Nature* 399, 429–36.

PINKER, S. 1994: *The Language Instinct: How the Mind Creates Language* (New York, William Morrow).

PRICE, T.D. and BROWN, J.A. 1985: *Prehistoric Hunter-Gatherers: the Emergence of Cultural Complexity* (Orlando, Academic Press).

PUTNAM, R.D. 1993: *Making Democracy Work: Civic Traditions in Modern Italy* (Princeton, Princeton University Press).

RICHERSON, P.J. & BOYD, R. 1989: The role of evolved predispositions in cultural evolution: or human sociobiology meets Pascal's wager. *Ethology and Sociobiology* 10, 195–219.

RICHERSON, P.J. and BOYD, R. 1998: The evolution of human ultra-sociality. In Eibl-Eibesfeldt, I. & Salter, F.K. (eds.), *Indoctrinability, Ideology, and Warfare* (New York, Berghahn).

RICHERSON, P.J. & BOYD, R. 1999: Complex societies: the evolutionary origins of a crude superorganism. *Human Nature* 10, 253–89.

RICHERSON, P.J. & BOYD, R. 2000: Built for speed: Pleistocene climate variation and the origin of human culture. *Perspectives in Ethology* 13, 1–45.

RICHERSON, P.J. & Boyd, R. in press: The evolution of subjective commitment groups: a tribal social instincts hypothesis. In Nesse, R.M. (ed.), *The Evolution of Subjective Commitment* (New York, Russell Sage Foundation).

RICHERSON, P.J., BOYD, R., & BETTINGER, R.L. in press: Was agriculture impossible during the Pleistocene, but mandatory during the Holocene? A climate change hypothesis. *American Antiquity*.

RINDOS, D. 1984: *The Origins of Agriculture: an Evolutionary Perspective* (London, Academic Press).

ROGERS, E.M. 1983: *Diffusion of Innovations*, 3rd edn (New York, Free Press).

ROOF, W.C., & McKINNEY, W. 1987: *American Mainline Religion: Its Changing Shape and Future* (New Brunswick, NJ, Rutgers University Press).

SAHLINS, M. 1963: Poor man, rich man, big-man, chief: political types in Melanesia and Polynesia. *Comparative Studies in Sociology and History* 5, 285–303.

SALTER, F.K. 1995: *Emotions in Command: a Naturalistic Study of Institutional Dominance* (Oxford, Oxford University Press).

SEARS, D.O. & FUNK, C.L. 1990: Self interest in Americans' political opinions. In Mansbridge, J.J. (ed.), *Beyond Self-Interest* (Chicago, University of Chicago Press), 147–70.

SERVICE, E.R. 1975: *Origins of the State and Civilization* (New York, Norton).

SIMON, H.A. 1990: A mechanism for social selection and successful altruism. *Science* 250, 1665–8.

SMITH, B.D. 1995: *The Emergence of Agriculture* (New York, Freeman).

SQUIRES, A.M. 1986: *The Tender Ship: Government Management of Technological Change* (Boston, Birkenhäuser).

SRINIVAS, M.N. 1962: *Caste in Modern India* (Bombay, Asia Publishing House).

STEWARD, J.H. 1955: *Theory of Culture Change: the Methodology of Multilinear Evolution* (Urbana, University of Illinois Press).

SUGDEN, R. 1986. *The Economics of Rights, Co-operation, and Welfare* (Oxford, Blackwell).

SYMONS, D. 1989: A critique of Darwinian Anthropology. *Ethology and Sociobiology* 10, 131–44.

TOMASELLO, M. 1999. *The Cultural Origins of Human Cognition* (Cambridge, MA, Harvard University Press).

TULLOCK, G. 1965: *The Politics of Bureaucracy* (Washington DC, Public Affairs Press).

TURNER, J.H. 1995: *Macrodynamics: Toward a Theory on the Organization of Human Populations* (New Brunswick, NJ, Rutgers University Press).

VAN DEN BERGHE, P.L. 1981: *The Ethnic Phenomenon* (New York, Elsevier).

WEST, R. 1941: *Black Lamb and Grey Falcon: A Journey Through Yugoslavia* (New York, Viking).

WIESSNER, P. 1984. Reconsidering the behavioural basis for style: a case study among the Kalihari San. *Journal of Anthropological Archaeology* 3, 190–234.

WIESSNER, P. 1996: Levelling the hunter: constraints on the status quest in foraging societies. In Wiessner P. & Schiefenhövel W. (eds.), *Food and the Status Quest* (New York, Berghahn), 171–91.

WIESSNER, P. & TUMU, A. 1998: *Historical Vines: Enga Networks of Exchange, Ritual, and Warfare in Papua New Guinea* (Washington DC, Smithsonian Institution Press).

WILLIAMS, G.C. 1966: *Adaptation and Natural Selection* (Princeton, Princeton University Press).

WILSON, E.O. 1975: *Sociobiology: A New Synthesis* (Cambridge, MA, Harvard University Press).

WINTERHALDER, B. 1986: Diet choice, risk, and food sharing in a stochastic environment. *Journal of Anthropological Archaeology* 5, 369–92.

WINTERHALDER, B. & GOLAND, C. 1997: An evolutionary ecology perspective on diet choice, risk, and plant domestication. In Gremillion K.J. (ed.), *People, Plants, and Landscapes: Studies in Palaeoethnobotany* (Tuscaloosa, University of Alabama Press), 123 60.

ZOHARY, D. & HOPF, M. 1993: *Domestication of Plants in the Old World: the Origin and Spread of Cultivated Plants in West Asia, Europe, and the Nile Valley*, 2nd edn (Oxford, Oxford University Press).

From Nature to Culture, from Culture to Society

W. G. RUNCIMAN

INTRODUCTION

OVER THE PAST TWENTY YEARS, neo-Darwinian theory has brought about a fundamental change in the way that the concept of evolution is defined and applied in the social no less than in the natural sciences. It is not just that the macrosociological teleology common to Marxism and Social Darwinism alike (Runciman 1989) has been finally jettisoned, as has the unilinear neo-evolutionary anthropological theory of the 1950s and '60s (Shennan 1999; Yoffee 1993), but that Darwin's original insight about 'descent with modification' — or, as it is nowadays put, 'heritable variation and competitive selection' — has come to be widely recognized, including by archaeologists (Maschner 1996; Spencer 1990: 4), as the general paradigm for non-teleological explanation of qualitative change. I have discussed elsewhere some of the reasons why cultural and social evolution need to be clearly distinguished and separately analysed (Runciman 2001). In this chapter, I assume without further argument that cultural evolution is both analytically and historically prior to social evolution, and consider the theoretical implications which follow for the explanation of the origin of social institutions. In so doing, I draw on some of the ethnographic evidence on hunters and foragers which can, with due caution, be used in the attempt to reconstruct the behaviour-patterns of human beings between the Upper Palaeolithic and Neolithic 'Revolutions' (if such they were), and also on game-theoretic studies that can help to elucidate the co-operative relationships extending beyond kinship which can perhaps be inferred from the evidence of the archaeological record. Inevitably, I rely on the work of authors whose findings and interpretations I have no competence to dispute. But that is the normal predicament of all comparative and historical sociologists.

Proceedings of the British Academy, **110**, 235–54, © The British Academy 2001.

THE TWO TRANSITIONS

In cultural selection, in contrast to natural selection, the transmission of information affecting phenotype is from the mind of one organism to the mind of another organism by teaching or imitation (Boyd & Richerson 1985: 33). But despite the obvious differences between biological and cultural evolution, selective pressure acts in both cases on the more or less extended effects of heritable instructions affecting phenotype in such a way as either to enhance or to diminish the likelihood of the continuing replication and diffusion of those instructions. The term 'meme', although used elsewhere in this volume without commentary, is perhaps best avoided until there is less disagreement than at present about its precise definition (Blackmore 1999: 63–6). But, however and wherever encoded and stored (Lake 1998), and whether deontic or merely permissive in logical form, bundles of instructions affecting phenotype are constantly being transmitted to adjacent or successive populations by imitation or learning (or both). The mutations or recombinations which occur on transmission may (and often do) arise from active and conscious reinterpretation by the receiving minds, but their continuing replication and diffusion still depend on whether they are fit and hence selected.

So defined, 'culture' is not unique to humans, and long-term field studies of chimpanzees, in particular, have disclosed a far wider range of variations transmitted by genuine imitation and learning, as opposed to stimulus enhancement or operant conditioning, than previously supposed (Whiten *et al.* 1999). But only humans have a capacity for sustained and cumulative cultural evolution (Boyd & Richerson 1996), and once language was fully developed both the range and the speed of potential variation could not but be dramatically enhanced. For the purpose of this chapter, it does not matter when or how the capacity for grammar and syntax evolved. It is enough that it was a contingently sufficient condition of the social behaviour-patterns of the Middle to Upper Palaeolithic transition (Mellars 1996; Mellars & Stringer 1989).

These behaviour-patterns, however, are explicable without any reference to the institutional rules and associated inducements and sanctions which define the modes of production, persuasion, and coercion of the societies of the Neolithic and thereafter. It was not until human beings began to lead their lives in an emergent world of armies, markets, temples, estates, treasuries, assemblies, courts (in both senses), schools, officials (public or private), taxes (or tribute), and inherited differences in status that the instructions affecting phenotype came to be formulated in rules which for the first time explicitly prescribed the reciprocal behaviour of pairs of agents who, whatever their individual differences, now behaved towards each other as nobles and commoners, landlords and tenants, masters and slaves, rulers and subjects, priests and laity, judges and plaintiffs, and so forth. Except at rare times of constitutional choice, individual agents such as these have no say over the rules which make them what they are: they grow up to find them already encoded in the practices constitutive of their society's economic, ideological, and political institutions. But heritable information affecting phenotype is once again being replicated and diffused to the extent that it is fit and hence selected, and the explanation of social, as previously of cultural, evolution depends on identifying the extended phenotypic effects of the bundles of instructions on which specific features of the environment bring selective pressure to bear.

The nature of the difference between natural, cultural, and social selection emerges distinctly from the different uses of the term 'role'. In the behaviour-patterns of animals, the instinctive capacity for mutual recognition of roles such as 'occupier' and 'intruder' is evident (Maynard Smith 1982: 204), and genetically transmitted instructions determine how individuals will respond in consequence of that recognition. In cultural selection, instinct is now supplemented by imitation and learning, and roles such as 'leader' or 'enemy', even if behaviour towards them may still be 'evoked' as well as acquired (Tooby & Cosmides 1992: 116–18), are not just biologically defined: individuals are chosen on the basis of personal attributes either as role-models to be admired or, on the contrary, as members of out-groups to be stigmatized. In social selection, roles are, as in the examples given above, defined by practices whose rules govern the reciprocal behaviour of both parties in a relationship underwritten by inducements and sanctions which may not always be effective in controlling the behaviour in every individual case, but still sustain the economic, ideological, or political institutions constituted by the roles carrying the practices. Furthermore, a social role can maintain its institutional existence even if no individual agent is occupying it at a given time, and social roles are at the same time parts to be performed in culturally variable ways and, in structural terms, vectors in a three-dimensional social space corresponding to the axes of economic, ideological, and coercive power. It is true that power is involved in natural and cultural as well as in social selection. Animals contesting for food or territory or access to mates are clearly engaged in a power struggle, as are preachers or prophets seeking to recruit adherents to their rival systems of value and belief. But in social selection, the inducements and sanctions are no longer interpersonal only.

In both transitions, from natural to cultural selection and from cultural to social, there must have been an originating event, just as in the evolution of the human species through natural selection there must have been a 'mitochondrial Eve'. Somebody must have been the first person to think of depicting something they had seen by drawing it on the wall of a cave (or elsewhere) and thus become the first representational artist, and some pair of persons must have been the first to exchange labour for payment on a contractual basis and thus to bring wage-labour into being. Although there is no way of finding out where

W. G. Runciman

and when, it does not follow that the critical mutations which turned out to have a significant effect on the subsequent course of cultural and social evolution are to be treated as random. In both cultural and social selection, mutations are likely to be both conscious and deliberate. But that is no guarantee of their success. Social, no less than natural, scientists have to be wary of falling into the 'Genetic Fallacy' of assuming that to ascertain the cause of a novel event is thereby to explain its consequences.

CULTURE WITHOUT SOCIETY

The anatomically modern humans of the Upper Palaeolithic, however significant their differences from chimpanzees or from other hominid species, shared with them an inherited 'social' or 'Machiavellian' intelligence (Byrne & Whiten 1988; Whiten & Byrne 1997) and a disposition, shared with other apes and also with monkeys (de Waal 1996: 102), to maintain tolerably stable dominance orders while at the same time probing them for weaknesses and opportunities for advancement. This raises two related sociological questions about the human behaviour-patterns of the Upper Palaeolithic. First, why did they not more quickly evolve from kin-based, small-group interpersonal relationships into the social institutions which appear only many millennia later? Second, how, without such institutions, but with an inherited propensity to compete for dominance, did they remain as stable as they did?

The second question leads in turn to the further question as to how ongoing co-operative relationships can extend beyond kin-groups (including, it may be, adoptive or fictive kin) within which the incidence of altruistic behaviour can be predicted to follow Hamilton's Rule. Despite the claims made by or on behalf of Axelrod (1984) for tit-for-tat as an evolutionarily stable strategy in indefinitely continuing iterations of the Prisoner's Dilemma, neither tit-for-tat nor any other pure strategy is uninvadable (Binmore 1998). But that is not to say that co-operation cannot be sustained in groups of unrelated individuals. The likelihood that co-operative strategies will be able to resist invasion by more than small numbers of free-riders is significantly increased by non-random association of co-operators with other co-operators (Kitcher 1993), by beliefs about the proportion of co-operators in the population (McKelvey & Palfrey 1992), by group discussion (Caporeal *et al.* 1989), by the restrictions imposed on free-riders by search time and coalition time (Enquist & Leimar 1993), and by sufficient frequency of interaction for future strategies to be based on the past behaviour of others (Cox *et al.* 1999). Furthermore, co-operators who are prepared to punish both free-riders and non-punishers (Boyd & Richerson 1992; Hirshleifer & Rasmusen 1989) can stabilize a behaviour-pattern within a population despite the problem that if punishment is costly to individuals selection might be expected to favour co-operators who decline to punish. More generally still, the disposition of conformists to act against non-conformists, if only through gossip, ostracism, or other forms of expression of moral disapproval, is too well known to need documentation, and the making and keeping of promises can be strengthened, although not guaranteed, by the mutual co-ordination and performance of ritual actions implying commitment to future co-operation (Watanabe & Smuts 1999: 101).

All this suggests that bands of hunters and foragers living under the environmental conditions of the Upper Palaeolithic could very well have maintained stable networks of co-operation extending beyond kinship without either a political sub-group imposing order, a religious sub-group maintaining conformity, or an economic sub-group controlling the distribution of resources. Such bands do not need to be strictly egalitarian in order to maintain their cohesion (Flanagan 1989). Not all their members will have had equal access to material goods or personal possessions, relative intellectual influence and prestige (Brunton 1989), or ability to inflict or resist physical violence, and such interpersonal differences can generate inequalities in individual life-chances just as great as those generated by systactic differences in the power attaching to institutional roles: unskilled band members whose performance of domestic tasks is controlled by their elders, unpopular band members denied participation in ceremonial events, or undisciplined band members singled out for physical punishment by their peers might as well, from their point of view, be subjected to the institutional domination of proprietors, priests, or police. But there will have been scope not only for coalitions of the kind observable in primate groups, but for the kind of long-term information-sharing not accounted for by kinship or reciprocal altruism which is observable in small human communities with a high degree of intertwined social relationships (Palmer 1991). There will always have been potential alpha-male 'aggrandizers' ready for opportunities to exploit 'prestige' technology and create for themselves the roles of an established, self-perpetuating elite (Hayden 1998). But there is, so far as I am aware, no evidence in the archaeological record to indicate that such opportunities were yet available to any significant degree, or that self-sustaining cycles of intergroup competition, production of surpluses for ritual feasting, expansion of social networks, and further intergroup competition (Lourandos 1988: 159) were yet under way. The sanctions deployed against 'aggrandizers' in numerous well-documented latter-day hunting and foraging bands (Boehm 1999; Sober & Wilson 1998: Chapter 5) may not be quite the same as the sanctions which were deployed in the Upper Palaeolithic (Binmore, this volume). But these were evidently effective unless and until there was a sufficiency of resources of the kind familiar from the well-documented aquatic societies of the north-west Pacific coast (Ames 1994) and elsewhere (Arnold 1996). In the words of Maschner and Patton (1996: 101), 'hereditary social status will develop everywhere the social and economic circumstances will allow it'. But in the Upper Palaeolithic, the socially stratified behaviour-patterns of the Neolithic are, so to speak, not yet within sight.

The difficulty of extrapolating from the archaeological record the states of mind of the persons whose behaviour created that record has long been recognized by archaeologists. But the psychologically (Jolly 1999; Mithen 1996), as well as anatomically, modern humans of the Upper Palaeolithic cannot but have talked to each other about illness, death, the elements, the heavenly bodies, and the behaviour of birds and animals as well as about food, shelter, and sex, and their culture, however much of it is now irrecoverable, was unmistakably more sophisticated than what had preceded it. Suppose, therefore, that their concomitant behaviour-patterns were as complex as a generous interpretation of the archaeological evidence allows. Would that require the transmission of instructions affecting phenotype by anything other than interpersonal imitation and learning, perhaps accelerated by the mechanisms of frequency-dependence and indirect bias (Bettinger 1991: Chapter 8)? There could have been population increase, division of labour, long-distance exchange, warfare, communal activities, and consistent performance of ritual without institutional inducements or sanctions being necessary.

Population increase

If more women than hitherto survived for long enough to bear more children and band sizes increased, this did not need to lead to an institutionalization of social relationships, simply because of what Soffer (1989: 722) calls 'the most powerful cause of egalitarian socio-political relationships among hunter-gatherers: the ability to vote with one's feet'. Band size may have been kept low also by infanticide, post-partum sexual taboos, and senilicide. But given relatively low overall rates of population growth, such as have been hypothesized for the Palaeolithic, and the availability of adjoining territory, fission was the available alternative to potentially unmanageable increases in band size.

Division of labour

The performance of different tasks by different members of hunting and foraging bands, starting with hunting by men and gathering by women, is well documented in the ethnographic record. But it could extend a good deal further without requiring any institutional inducements or sanctions: harvesting of wild cereals could have been done by interpersonal agreement without anything approaching formal employment relations, and skilled craftsmen working in bone or ivory who decorated hunting weapons with the likenesses of animals did not need to be under the control of dominant groups in the way that they were when 'craftsman' had become a specialized subordinate role of the kind familiar from palace and temple compounds. Nor did ritual specialists need to be the incumbents of distinctive institutionalized roles any more than individuals particularly skilled in hunting or persuasive in group discussion: divination, or prediction of the weather, or diagnosis of sickness, or choice of camp-sites or hunting-grounds could all have been done on a personal basis by whichever member of the band was informally agreed to do them best.

Long-distance exchange

The exchange of material artefacts between bands over long distances in the Upper Palaeolithic seems agreed among archaeologists to have functioned less as trade in an economic sense than as information exchange integrating bands which needed alliances for mating and for insurance against resource shortfalls (Gamble 1982). This was not the sort of long-distance traffic organized, for example, by the Maghribi traders of medieval Cairo through informal coalitions which functioned also as an information-transmission mechanism (Greif 1989). Nor was it the sort of 'diplomatic' exchange of high-prestige gifts between high-status 'guest-friends' familiar from post-Mycenaean Greece (Snodgrass 1980: 55–6). Symbolically valued materials, including the use of red ochre for body decoration, could, like linguistic codes (Nettle & Dunbar 1997), function as cultural markers differentiating one from another group without thereby giving rise to institutionalized roles.

Warfare

Lethal violence appears in the archaeological record from the earliest-dated burials of men, women, and children whose wounds, including wounds inflicted by projectile points, can only be explained as resulting from deliberate assault (Carman & Harding 1999; Keeley 1996: 37–8). If 'warfare' is defined in terms of specialized military institutions, professional generals, protracted campaigning, and formal chains of command, then there is no evidence for warfare in the Upper Palaeolithic. But lethal violence between mutually hostile groups requires nothing more than informal co-operation among males similar to the co-operation in chimpanzee groups raiding alien territory (Goodall 1986).

Communal activities

The use of designated public places for communal activity is unmistakably visible only when physically marked out in such a way that it becomes appropriate to apply to them terms such as agora, compound, precinct, arena, forum, plaza, stadium, or henge. By this time, the communal activities are likely to include trading of goods or services, mustering of citizens or soldiers, and celebration of rulers or divinities, which all presuppose established practices defining acknowledged social roles. But the bands of the Upper Palaeolithic could have assembled regularly and in numbers at designated places for such communal purposes as consulting, celebrating, observing the elements, sharing food, or simply exchanging news, telling stories, dancing, or playing games. Again, all these behaviour-patterns are cultural but need not also be institutional. Their different forms of expression can be defined and regulated by imitation and learning alone: for example, young men could perfectly well perform co-ordinated gymnastic or athletic displays of the kind put on for the benefit of Odysseus by Homer's Phaeacians (*Odyssey* VIII, 256–65) without, as in that case, being summoned by a chief (*basileus*) and organized by a herald (*kerux*) and umpires (*aisymnetai*). Or, to cite an ethnographic example, the young men who come to dance at a Nyakusa funeral may 'feel the need to confront death with an assertion of life' (Metcalf & Huntington 1979: 39) without any institutional inducements or sanctions being needed.

Performance of ritual

It may be no easier to reconstruct what was going on in the minds of the men and women of the Upper Palaeolithic from their burials than from their art (Parker Pearson 1999). But even if the contents of their graves are taken to imply religious beliefs and values of an elaborate and systematic kind, it does not follow that the transition from culture to society had been made. Considerable cultural variation is possible in the choice of material objects interred with the corpse, treatment of children relative to adults, or inclusion of domestic animals, without the kind of institutional role-structure unmistakably reflected in such grave goods as bronze armour, gold death-masks, signet-rings, or miniature iron toy chariots (Morris 1998: 44). Even where children are buried with what look like valuables, this may be testimony to the mourners' grief rather than the child's social status (Jacobs 1995). Similarly, even if rock art implies shamanistic claims to arcane knowledge (Lewis-Williams 1995) and initiation ceremonies not accessible to every male, let alone female, band member (Owens & Hayden 1997), we are still a long way from a Weberian priesthood with its own corporate *Machtstellung*.

In summary, therefore, the behaviour-patterns of psychologically modern, linguistically competent, technically skilled human groups in the Upper Palaeolithic could well have been complex and variable in content, while at the same time sufficiently stable in form for continuity over successive generations, without there being any formally defined and institutionally sanctioned economic, ideological, or political roles. No doubt it is possible that there were transi-

tional forms of proto-institutional social organization: sons may have succeeded fathers as prominent warriors, craftsmen, or shamans, vigorous leaders
may on occasion have mobilized substantial workforces for collective purposes,
large descent-groups may have sustained ongoing claims to selected ritual sites,
local 'big men' (Johnson & Earle 1989: 57) may have kept more women and distributed more food than other adult males, and some families may have
devoted more work-time to making elaborately decorated clothing or bodily
ornaments, or to collecting valued material objects. But the bundles of instructions affecting phenotype could still have been transmitted by interpersonal
imitation and learning without being encoded in practices defining institutional roles. Moreover, imitation and learning could quite well have accounted
by themselves for ongoing between-group, as well as within-group, co-operation
adequate for the maintenance and renewal of long-distance, long-term tribal
and inter-tribal relationships. Even 'super-networks' associated with extensive
exchange systems (Lourandos 1997: 26) require only culturally transmitted
greeting ceremonies and rites of entry (Peterson 1975) and interpersonally
transmitted recollections of past encounters with members of other bands.

SOCIETY OUT OF CULTURE

What then happened to bring about the transition from a purely cultural to an
institutional, as well as cultural, human world?

There is an extensive literature, both archaeological and ethnographic, on
states and chiefdoms and their origins. But although the evolution of proto-
states (Runciman 1982) and the subsequent diffusion of centralized coercive
practices can be very rapid, particularly in response to selective pressure from
other, already stratified societies, the first emergence of roles formally defined
by rule-governed practices is the critical event which, like subsequent state-
formation, evidently happened more than once in different parts of the world.
Much of the discussion in the literature on the evolution of social complexity
is concerned with the influence of the traditional 'prime movers': population
growth, trade, warfare, and religion. But perhaps only a minimal extra accumulation of resources and differentiation of functions is enough to initiate the
transition from culture to society. Not only does it not require 'chiefs' (however
defined); it may not even require the emergence of 'rank' — a concept which in
any case has imprecisions and ambiguities of its own (Renfrew 1982). Socio-
logically, the significant difference is that information affecting phenotype is
now encoded and transmitted in such a way that different individuals can move
into and out of, and be succeeded in by other individuals, ongoing roles whose
defining practices are acknowledged by mutually responsive agents independently of personal characteristics.

The theoretical point can be illustrated by simple hypothetical examples. Household heads with control over a relatively larger quantity of stored food-stuffs might, in times of scarcity, start making distributions to less fortunate households on terms which involved subsequent repayment in kind and hence a recognized practice formalized in a system of debtor–creditor relationships. Or ritual specialists credited with divinatory powers might build up a follow-ing, centred perhaps around a shrine or sacred place, from among whom designated successors would be chosen in accordance with rules defining a for-mal master–disciple relationship. Or particularly redoubtable warriors might attach to themselves a permanent retinue who thereby enabled them to rely on continuing support from fellow-warriors without repeated demonstrations of personal prowess. The emergence of such relationships may well (again like subsequent state-formation) have been fluid, sporadic, and reversible. Where exactly is the line to be drawn between exchanges of favours between friends and the mutual obligations of patrons and clients, or between gift-exchanges and bridewealth or dowry payments, or between informal teaching of skills and formal training for full-time, specialized functions, or between services rendered to immediate family members and labour diverted to local corporate groups sharing a putative common ancestor? Where ethnographic, literary, or epigraphic evidence is available, the emergence of novel practices can be inferred from the vernacular terms for the roles which they define, like the Alaskan *umialik*, whose role is argued by Sheehan (1985: 142) to have been transformed by whaling from successful hunter to war leader, wealthy trader, and religious leader; and conversely, the repudiation of an available term for a role, as in the reluctance of the !Kung San to apply their word for 'chief' to themselves as opposed to Bantu headmen except 'in a derisory manner' (Lee 1979: 344), can confirm the absence of institutional relationships. But the epigraphic or literary evidence for roles with distinctive names attaching to them, such as the Homeric Phaeacians' *basileus*, *kerux*, and *aisymnetai*, or the Sumerian *ensi*, or the Mycenaean *wanax*, is likely to come from a time when the origin of the practices defining them is already a long-past event.

One or more just-so stories about the evolution from culture to society must, however, be true; and what the archaeological record does suggest is that common to those which are will, at some point and to some degree, be seden-tism. This is not because permanent economic, ideological, and political institutions are bound to evolve from any kind of synoecism. Nor, in any case, is sedentism an all-or-nothing matter. Hunting and foraging bands can be vir-tually sedentary for significant periods and thereby make and acknowledge claims to territoriality (Cashdan 1980) without year-round occupation of caves, huts, or houses. Even where permanent stone-built structures have been uncovered, this does not necessarily indicate sedentism, and even intensive harvesting of wild cereals need not imply more than a seasonal mobility pattern (Bar-Yosef & Meadow 1995). But once climatic conditions favoured continuous plant cultivation, as they evidently did in the early Holocene (Bar-Yosef, this volume; Richerson & Boyd, this volume), sedentism and intensive exploitation of annual crops are likely to have reinforced one another (Henry 1989; McCorriston & Hole 1991), and communities where both plants and animals had been domesticated will have started to experience a joint expansion of food surpluses and population (Hole 1984).

Whatever time it may have taken, a combination of storage, sedentism, and residential aggregation could not but have had a significant effect on social behaviour-patterns, if only because of the need for co-operation within and between groups whose increased size made increasingly difficult if not impossible the detection of free-riders and restraint of aggrandizers. Human beings as a species appear to have evolved adaptive social networks limited to about 150 people (Dunbar 1993, 1998), and increasing group size is well known to give rise to problems of collective decision-making which need to be alleviated either by prescriptive rituals or by diffusion of decision-making among clans, moieties, sodalities, or age-grades (Johnson 1982; Reynolds 1984). But large aggregations and absorption of population from other groups also give rise to problems of social control (Kaufman 1992) and conflicts of loyalty (Myers 1988: 59) which are compounded to the extent that there are rival claimants to increasingly contested resources, including access to, and control of, physical space. The same co-operative norms may be transmitted by imitation and learning from parents and elders to successive generations of children, but how are they to be made effective in cases of inter-familial dispute? There do not need to be central, permanent, specialized roles to which there attaches monopoly control of the means of coercion. But some formal procedure is now called for whereby disputes can be resolved by mediators or arbitrators designated as such, in contrast to the purely personal interventions in fights among foragers like the !Kung San, where a man with a reputation for being strong and competent and himself very mild-mannered can successfully interpose himself, but women or old men who interpose themselves are often 'hit in the bargain' for their pains (Lee 1979: 308, 381). The formal roles of mediator or arbitrator pose the difficulty that the practices defining them could emerge without leaving any trace whatever in the archaeological record. But it is surely legitimate to conjecture that, for example, the small Natufian building in Ain Mallaha with a plaster-covered rounded bench 'could have been used by the leader or shaman of the group' (Bar-Yosef 1998: 163). This is not a ruler's throne, or a magistrate's judgement-seat, or an official's dais, or a chief's stool, any more than the caves with the paintings are churches or temples. But nor is it just the camp-fire or water-hole round which debate is conducted by the men and women of the hunting and foraging bands.

From such preliminary indications of emergent practices and roles, one possible trajectory leading to a stable equilibrium is what may be called the 'Deioces model'. According to Herodotus (I.96–100), Deioces, having acquired a reputation as a mediator, refused to continue until granted his own armed retainers and palace, and thus became the first king of the Medes. Although the particular story is mythical, many 'aggrandizers' in the historical record have succeeded in usurping, or being granted, a monopoly of the means of coercion, and from there it can be a short step to religious legitimation of rulership, honorific modes of address, slavery or forced labour, a military and perhaps landholding elite, and a hierarchical structure of designated roles whose incumbence is hereditary. But co-operation does not have to be imposed in this way even in large, settled communities with distributable surplus resources. The contrasting solution, in game-theoretic terms, is to add to the players' decision-tree a parameter representing a cost for self-financing monitoring and a strategy for negotiating a co-operative agreement (Ostrom 1990: 16). As Ostrom emphasizes (1990: 14), 'New institutional arrangements do not work in the field as they do in abstract models unless the models are well specified and the participants in a field setting understand how to make the new rules work.' But examples exist, and have been studied in the field, where, for example, fishermen agree to rotate potentially productive locations in such a way that access to the best opportunities is equalized and monitoring is carried out by mutual observation at sufficiently low individual cost for it not to be a rational strategy to defect. This is not to say that all such arrangements are successful, or that there are not sub-optimal polymorphic equilibrium traps into which the relevant population can fall (Skyrms 1996). But as one of the early contributors to the literature on common-pool resource problems remarked, 'stable primitive cultures appear to have discovered the dangers of common-property tenure and to have developed measures to protect their resources. Or, if a more Darwinian explanation be preferred, we may say that only those primitive cultures have evolved which succeeded in developing such institutions' (Gordon 1954: 134–5).

These two alternative models are close to being ideal types of the alternative 'Hobbesian' and 'Rousseauesque' versions of how control is exercised within and between groups whose members owe no familial or personal allegiance to people whose interests conflict with their own. But the ethnographic record discloses a whole range of intermediate arrangements in hunter-gatherer, as well as horticultural or agricultural or pastoral, societies by which the continuity of distinctive and consistent social behaviour-patterns can be sustained. It is a commonplace that no social system is based entirely on consent or entirely on compulsion. But in the diverse combinations of practices and roles which make up the institutions of relatively simple but none the less very different societies, there is a universal tendency for the phenotypic effects of economic, ideological, and political practices to function in such a way as to enhance the probability of replication of one another, often through recombination in a single composite role. It is not only in the formation of states that 'the elaboration of religious institutions, ideology, and the arts' is so frequently involved (Bettinger *et al.* 1996: 159). Even if there is no warrant for assuming that any of the hunter-forager societies in the ethnographic record is a replica of those of the Epi-Palaeolithic (Betzig 1998: 267; Binmore, this volume; Mithen 1994: 170; Shott 1992), the inference which can legitimately be drawn from the ethnography of the more affluent and complex hunter-gatherers is the ease with which environmental pressure and competition between the incumbents of differentiated roles can generate institutional as well as cultural variation once the evolutionary threshold has been crossed between information transfer by imitation and learning only and the encoding of formal rules in the mutant and recombinant practices which define these differentiated roles.

This is what makes the first completion of the transition of such obvious sociological interest and gives the Natufian culture its particular relevance (Bar-Yosef, this volume). I have suggested that even on a generous interpretation of the archaeological evidence, it is difficult to argue that the Upper Palaeolithic 'Revolution' brought the transition about. But even on a sceptical interpretation of the Natufian evidence, it is difficult to question that it has occurred. 'Transition', however, rather than 'revolution', would seem to be the appropriate word. Subsequent social evolution can be very rapid because, among other reasons, the idea of institutional inducements and sanctions has been arrived at already, and the memory or suggestion of formally differentiated roles can be culturally transmitted from one successive or adjacent population to another. But when the first mutant practices brought the first economic, ideological, or political institutions into being, the vernacular terms for the roles defined by those practices will have had to be invented or new meanings attached to existing terms. What is more, not every member of the community in question will necessarily have given their assent. Nor, as the Natufian example also shows, is there any reason to assume that the transition may not go into reverse (Bar-Yosef, this volume; Moore & Hillman 1992).

Successive and continuing excavations have by now yielded a detailed and coherent picture of Natufian settlements which, although significantly smaller than Neolithic sites such as Gilgal or Jericho, were evidently as large as the smaller Neolithic villages. Despite the relative rarity of underground storage pits, there appears to be unmistakable evidence of delayed-return surpluses, albeit not on a scale sufficient for Kwakiutl-style potlatching. Even if cereal crops were not yet fully domesticated (Unger-Hamilton 1989: 101), plant remains, together with sophisticated tools and extensive faunal remains, indicate skilled and productive exploitation of the natural environment. An incipient industry produced decorated pendants and beads as well as tools and parts of hunting devices such as spear-throwers. Grave goods include belts, bracelets, earrings, and headpieces. Skulls were preserved, perhaps as symbols of deceased family members, or perhaps as trophies of dead enemies (Bienert 1991: 20). Marine shells, basalt implements, and obsidian pieces were acquired from sources distant from the locations where they have been found. Artworks in distinctive styles include decorated bowls, spatulas, and sickles, human and animal figurines, and incised limestone slabs, some of which have what may be notational marks. Could all these activities have been carried out by communities in which there were no acknowledged practices defining specialized formal roles in which different individuals succeeded one another according to pre-existing rules? Must there not have been what have been called in the Mesoamerican context 'central planners of seasonal working performance' (Boehm de Lamenas 1988: 93) who, although they need not have been 'chiefs', let alone 'kings', were still recognized as formally entitled to direct the labour of people other than their own families and households, at least to some degree? Perhaps systems of stably integrated, explicitly designated, formally acknowledged, extra-familial economic, ideological, and political roles emerged only in later, larger, fully sedentary communities with public spaces, communal buildings, monuments, domesticated crops, trading networks, and formal ceremonials. But these communities evolved from the Natufian communities which preceded them: their members were 'the descendants of the local Natufian population which had undergone changes in material culture, social organization, and daily life ways' (Bar-Yosef 1998: 169).

Later autonomous transitions, whether in Mesoamerica or the Yangtze Valley, need not have come about through an evolutionary sequence precisely following that suggested by the archaeological record for south-west Asia. But whenever it happened, aggregation of population beyond a certain critical point created an environment favouring formal inducements and sanctions without which the necessary minimum of co-operation between unrelated individuals would no longer be possible to sustain. Subject to the risk attendant on any categorical statement of this kind, it would, I submit, be a sociological impossibility for the 10-hectare settlements of over 1,000 people of Pre-Pottery Neolithic B to have cohered and reproduced themselves without an institutionalized role-structure of a kind which the hunting and foraging bands of the Upper Palaeolithic did not need to have evolved in order to maintain their coherence over successive generations.

CONCLUSION

It may never be possible to reconstruct from the archaeological record exactly how there first evolved practices defining specific economic, ideological, and political roles, and still less what the vernacular terms were for them. In the ethnographic literature, there are numerous relatively simple societies with roles whose vernacular terms are suggestive of a possible origin. For example, the Achuan *juunt* (literally, 'great man') 'is invested with a pre-eminent role' because he is 'master of the house' and also because 'everybody relies on his recognized qualities of military leadership' (Descola 1996: 290); he therefore becomes the *mesetan chicharu* (literally, 'war herald') who both organizes defence and negotiates with allies or, if necessary, enemies. By contrast, the Etoro *tafidilo*, although likewise a leader in communal activities including raids against other tribes, is a respected senior man whose status is achieved 'by attaining prestige through generosity in the frequent distribution of growth-inducing game to co-residents' (Kelly 1993: 21), while the Mardujarra *nindibuga* or 'knowledgeable one' has a position 'based on the older men's monopoly of esoteric knowledge, which will be transmitted only if young men conform to the dictates of the Law, and are willing to hunt meat in reciprocal payment for the major secrets that are progressively being revealed to them' (Tonkinson 1988: 157). But any historical reconstruction can only be speculative.

On the other hand, the diversity of hunter-forager cultures (Kelly 1995) gives reason to suppose that there is a range of alternative ways in which the transition can come about, depending on the particular circumstances under which relatively larger populations stay together and settle more or less permanently at sites where they build up delayed-return resources. There is no suggestion that culturally complex but socially pre-institutional human behaviour-patterns were acted out in a pre-lapsarian idyll. The archaeological record discloses ample evidence of harsh climates, hostile predators, short life-spans, dietary deficiencies, incurable ailments, periodic scarcity, and interpersonal violence. But the selective pressures imposed on psychologically modern humans by a difficult, unpredictable, and often threatening environment will have favoured the replication of any of a range of possible mutations which furthered more consistent co-operation between unrelated members of larger groups and more stable relationships with other groups, whether perceived as competitors for scarce resources, sources of potential breeding partners, or links in a chain of long-distance exchange of symbolically valued material goods. To a comparative and historical sociologist it is as puzzling that any anthropologist should ever have questioned that hunters and gatherers 'do not all possess the same "ethos"' (Gibson 1988: 165) as that any archaeologist should ever have questioned that culture 'springs from the active engagement of people in the business of living and interacting' (Gamble 1999: 28; Whittle, this volume). The hunters and gatherers of the Upper Palaeolithic cannot but have consciously and deliberately constructed and negotiated their own particular relationships with the natural world, their artefacts and monuments (Renfrew, this volume), their deities, and each other, and out of this process some mutations in bundles of instructions affecting phenotype will have been more successfully replicated than others. Only, however, after a threshold had been crossed in sedentism, group size, and storage of surplus resources did they form and sustain institutions which depended for the maintenance of stable social relationships on obedience to rules encoded in established practices and formal roles acknowledged as such. If it took the Upper Palaeolithic 'Revolution' to complete the transition from 'evoked' to 'acquired' social behaviour, then perhaps it took the Neolithic 'Revolution' to complete the transition from 'acquired' social behaviour to 'imposed' (Runciman 1998: 174).

Note. I am grateful to Stephen Shennan for his comments on an earlier draft, but he bears no responsibility for the deficiencies which remain.

REFERENCES

AMES, K. 1994: The Northwest Coast: complex hunter-gatherers, ecology, and social evolution. *Annual Review of Anthopology* 23, 209–29.

ARNOLD, J.E. 1996: Organizational transformations: power and labor among complex hunter-gatherers and other intermediate societies. In Arnold, J.E. (ed.), *Emergent Complexity. The Evolution of Intermediate Societies* (Ann Arbor, International Monographs in Prehistory).

AXELROD, R. 1984: *The Evolution of Cooperation* (New York, Basic Books).

BAR-YOSEF, O. 1998: The Natufian culture in the Levant, threshold to the origins of agriculture. *Evolutionary Anthropology* 6, 157–77.

BAR YOSEF, O. & MEADOW, R.H. 1995: The origins of agriculture in the Near East. In Price, T.D. & Gebauer, A.B. (eds.), *Last Hunters, First Farmers: New Perspectives on the Prehistoric Transition to Agriculture* (Santa Fe, NM, School of American Research Press).

BETTINGER, R.L. 1991: *Hunter-Gatherers: Archaeological and Evolutionary Theory* (New York, Plenum).

BETTINGER, R.L., BOYD, R. & RICHERSON, P. 1996: Style, function and cultural evolutionary processes. In Maschner, H.D.G. (ed.), *Darwinian Archaeologies* (New York, Plenum).

BETZIG, L. 1998: Not whether to count babies, but which. In Crawford, C. & Krebs, D.L. (eds.), *Handbook of Evolutionary Psychology* (Mahwah, NJ, Erlbaum).

BIENERT, A.D. 1991: Skull cult in the prehistoric Near East. *Journal of Prehistoric Religion* 5, 9–23.

BINMORE, K. 1998: *Just Playing: Game Theory and the Social Contract II* (Cambridge, MA, MIT Press).

BLACKMORE, S. 1999: *The Meme Machine* (Oxford, Oxford University Press).

BOEHM, C. 1999: *Hierarchy in the Forest. The Evolution of Egalitarian Behavior* (Cambridge, MA, Harvard University Press).

BOEHM DE LAMENAS, B. 1988: Subsistence, social control of resources, and the development of complex society in the Valley of Mexico. In Gledhill, J., Bender, B. & Larsen, M.T. (eds.), *State and Society: The Emergence and Development of Social Hierarchy and Political Centralization* (London, Unwin Hyman).

BOYD, R. & RICHERSON, P.J. 1985: *Culture and the Evolutionary Process* (Chicago, University of Chicago Press).

BOYD, R. & RICHERSON, P.J. 1992: Punishment allows the evolution of cooperation (or anything else) in sizeable groups. *Ethology and Sociobiology* 13, 171–95.

BOYD, R. & RICHERSON, P.J. 1996: Why culture is common but cultural evolution is rare. *Proceedings of the British Academy* 88, 77–93.

BRUNTON, R. 1989: The cultural instability of egalitarian societies. *Man* n.s. 24, 673–81.

BYRNE, R.W. & WHITEN, A. 1988: *Machiavellian Intelligence: Social Expertise and the Evolution of Intelligence in Monkeys, Apes, and Humans* (Oxford, Clarendon Press).

CAPOREAL, L.R., DAWES, R.M., ORBELL, J.M. & VAN DE KRAGT, A.J.C. 1989: Selfishness examined: cooperation in the absence of egoistic incentives. *Behavioral and Brain Sciences* 12, 683–739.

CARMAN, J. & HARDING, A. (eds.) 1999: *Ancient Warfare: Archaeological Perspectives* (Stroud, Sutton).

CASHDAN, E.A. 1980: Egalitarianism among hunters and gatherers. *American Anthropologist* 82, 116–20.

COX, S.J., SLUCKIN, T.J. & STEELE, J. 1999: Group size, memory, and interaction rate in the evolution of cooperation. *Current Anthropology* 40, 369–77.

DESCOLA, P. 1996: *The Spears of Twilight: Life and Death in the Amazon Jungle* (London, HarperCollins).

DUNBAR, R.I.M. 1993: Coevolution of neocortical size, group size, and language. *Behavioral and Brain Sciences* 16, 681–94.

DUNBAR, R.I.M. 1998: The social brain hypothesis. *Evolutionary Anthropology* 6, 178–90.

ENQUIST, M. & LEIMAR, O. 1993: The evolution of cooperation in mobile organisms. *Animal Behaviour* 45, 747–57.

FLANAGAN, J.G. 1989: Hierarchy in simple 'egalitarian' societies. *Annual Review of Anthropology* 18, 245–66.

GAMBLE, C. 1982: Interaction and alliance in Palaeolithic society. *Man* n.s. 17, 92–107.

GAMBLE, C. 1999: *The Palaeolithic Societies of Europe* (Cambridge, Cambridge University Press).

GIBSON, T. 1988: Meat sharing as a political ritual: forms of transaction versus modes of subsistence. In Ingold, T., Riches, D. & Woodburn, J. (eds.), *Hunters and Gatherers II: Property, Power and Ideology* (Oxford, Berg).

GOODALL, J. 1986: *The Chimpanzees of Gombe* (Cambridge, MA, Harvard University Press).

GORDON, H.S. 1954: The economic theory of a common-property resource: the fishery. *Journal of Political Economy* 62, 124–42.

GREIF, A. 1989: Reputation and coalitions in medieval trade. *Journal of Economic History* 49, 857–82.

HAYDEN, B. 1998: Practical and prestige technologies: the evolution of material systems. *Journal of Archaeological Method and Theory* 5, 1–55.

HENRY, D.O. 1989: *From Foraging to Agriculture: the Levant at the End of the Ice Age* (Philadelphia, University of Pennsylvania Press).

HIRSHLEIFER, D. & RASMUSEN, E. 1989: Cooperation in a repeated prisoner's dilemma with ostracism. *Journal of Economic Behavior and Organization* 12, 87–106.

HOLE, F. 1984: A reassessment of the Neolithic Revolution. *Paléorient* 101, 49–60.

JACOBS, K. 1995: Returning to Oleni 'ostrol: social, economic, and skeletal dimensions of a boreal forest Mesolithic cemetery. *Journal of Anthropological Archaeology* 14, 359–403.

JOHNSON, A. & EARLE, J. 1987: *The Evolution of Human Societies: from Foraging Group to Agrarian State* (Stanford, Stanford University Press).

JOHNSON, G.A. 1982: Organizational structure and scalar stress. In Renfrew, C., Rowlands, R.J. & Segraves, B.A. (eds.), *Theory and Explanation in Archaeology: the Southampton Conference* (New York, Academic Press).

JOLLY, A. 1999: *Lucy's Legacy. Sex and Intelligence in Human Evolution* (Cambridge, MA, Harvard University Press).

KAUFMAN, D. 1992: Hunter-gatherers of the Levantine Epipalaeolithic: the sociocultural origins of sedentism. *Journal of Mediterranean Archaeology* 5, 165–201.

KEELEY, L.N. 1996: *War Before Civilization* (Oxford, Oxford University Press).

KELLY, R.C. 1993: *Constructing Inequality: the Fabrication of a Hierarchy of Virtue among the Etoro* (Ann Arbor, University of Michigan Press).

KELLY R.L. 1995: *The Foraging Spectrum: Diversity in Hunter-Gatherer Lifeways* (Washington, DC, Smithsonian Institution Press).

KITCHER, P. 1993: The evolution of human altruism. *Journal of Philosophy* 90, 497–516.

LAKE, M. 1998: Digging for memes: the role of material objects in cultural evolution. In Renfrew, C. & Scarre, C. (eds.), *Cognition and Material Culture: the Archaeology of Symbolic Storage* (Cambridge, McDonald Institute).

LEE, R.B. 1979: *The !Kung San. Men, Women, and Work in a Foraging Society* (Cambridge, Cambridge University Press).

LEWIS-WILLIAMS, J.D. 1995: Seeing and constructing: the making and 'meaning' of a Southern African rock art motif. *Cambridge Archaeological Journal* 5, 3–23.

LOURANDOS, H. 1988: Palaeopolitics: resource intensification in Aboriginal Australia and Papua New Guinea. In Ingold, T., Riches, D. & Woodburn, J. (eds.), *Hunters and Gatherers I: History, Evolution, and Social Change* (Oxford, Berg).

LOURANDOS, H. 1997: *Continent of Hunter-Gatherers. New Perspectives on Australian Pre-history* (Cambridge, Cambridge University Press).

McCORRISTON, J. & HOLE, F. 1991: The ecology of seasonal stress and the origins of agriculture in the Near East. *American Anthropologist* 93, 46–69.

McKELVEY, R. & PALFREY, T. 1992: An experimental study of the centipede game. *Econometrica* 60, 803–36.

MASCHNER, H.D.G. (ed.) 1996: *Darwinian Archaeologies* (New York, Plenum).

MASCHNER, H.D.G. & PATTON, J.Q. 1996: Kin selection and the origins of hereditary social inequality: a case study from the Northern Northwest Coast. In Maschner, H.D.G. (ed.), *Darwinian Archaeologies* (New York, Plenum).

MAYNARD SMITH, J. 1982: *Evolution and the Theory of Games* (Cambridge, Cambridge University Press).

MELLARS, P. 1996: The emergence of modern populations in Europe: a social and cognitive 'revolution'? *Proceedings of the British Academy* 88, 179–203.

MELLARS, P. & STRINGER, C. (eds.) 1989: *The Human Revolution: Behavioural and Biological Perspectives on the Origins of Modern Humans* (Edinburgh, Edinburgh University Press).

METCALF, P. & HUNTINGTON, R. 1979: *Celebrations of Death: the Anthropology of Mortuary Ritual* (Cambridge, Cambridge University Press).

MITHEN, S. 1994: Simulating prehistoric hunter-gatherer societies. In Gilbert, N. & Doran, J. (eds.), *Simulating Societies. The Computer Simulation of Social Phenomena* (London, University College Press).

MITHEN, S. 1996: *The Prehistory of the Mind: a Search for the Origins of Art, Religion and Science* (London, Thames and Hudson).

MOORE, A.M.T. & HILLMAN, G.C. 1992: The Pleistocene–Holocene transition and the human economy in Southeast Asia: the impact of the Younger Dryas. *American Antiquity* 57, 482–94.

MORRIS, I. 1998: Archaeology and archaic Greek history. In Fisher, N. & van Wees, H. (eds.), *Archaic Greece: New Approaches and New Evidence* (London, Duckworth).

MYERS, F. 1988: Burning the truck and holding the country: property, time and the negotiation of identity among Pintupi Aborigines. In Ingold, T., Riches, D. & Woodburn, J. (eds.), *Hunters and Gatherers II: Property, Power, and Ideology* (Oxford, Berg).

NETTLE, D. & DUNBAR, R.I.M. 1997: Social markers and the evolution of reciprocal exchange. *Current Anthropology* 38, 93–9.

OSTROM, E. 1990: *Governing the Commons. The Evolution of Institutions for Collective Action* (Cambridge, Cambridge University Press).

OWENS, D.A. & HAYDEN, B. 1997: Prehistoric rites of passage: a comparative study of trans-egalitarian hunter-gatherers. *Journal of Anthropological Archaeology* 16, 121–61.

PALMER, C.T. 1991: Kin-selection, reciprocal altruism, and information sharing among Maine lobstermen. *Ethology and Sociobiology* 12, 221–35.

PARKER PEARSON, M. 1999: *The Archaeology of Death and Burial* (Stroud, Sutton).

PETERSON, N. 1975: Hunter-gatherer territoriality: the perspective from Australia. *American Anthropologist* 77, 53–68.

RENFREW, C. 1982: Socio-economic change in ranked societies. In Renfrew, C. & Shennan, S. (eds.), *Ranking, Resource and Exchange* (Cambridge, Cambridge University Press).

REYNOLDS, R.G. 1984: A computational model of hierarchical decision systems. *Journal of Anthropological Archaeology* 3, 159–89.

RUNCIMAN, W.G. 1982: Origins of states: the case of Archaic Greece. *Comparative Studies in Society and History* 24, 351–77.

RUNCIMAN, W.G. 1989: Evolution in sociology. In Grafen, A. (ed.), *Evolution and its Influence* (Oxford, Clarendon Press).

RUNCIMAN, W.G. 1998: The selectionist paradigm and its implications for sociology. *Sociology* 32, 163–88.

RUNCIMAN W.G. 2001: Heritable variation and competitive selection as the mechanism of sociocultural evolution. In Boden, M., Ziman, J. & Wheeler, M. (eds.), *The Evolution of Cultural Entities* (Oxford, Oxford University Press).

SHEEHAN, G.W. 1985: Whaling as an organizing focus in Northwestern Alaskan Eskimo society. In Price, T.D. & Brown, J.A. (eds.), *Prehistoric Hunter-Gatherers* (Orlando, Academic Press).

SHENNAN, S.J. 1999: The development of rank societies. In Barker, G. (ed.), *The Companion Encyclopaedia of Archaeology* (London, Routledge).

SHOTT, M. 1992: On recent trends in the anthropology of foragers: Kalahari revisionism and its archaeological implications. *Man* n.s. 27, 843–71.

SKYRMS, B. 1996: *Evolution of the Social Contract* (Cambridge, Cambridge University Press).

SNODGRASS, A. 1980: *Archaic Greece* (London, Dent).

SOBER, E. & WILSON, D.S. 1998: *Unto Others: the Evolution and Psychology of Unselfish Behavior* (Cambridge, MA, Harvard University Press).

SOFFER, O. 1989: Storage, sedentism, and the Eurasian Palaeolithic record. *Antiquity* 63, 719–32.

SPENCER, C.S. 1990: On the tempo and mode of state formation: neo-evolutionism reconsidered. *Journal of Anthropological Archaeology* 9, 1–30.

TONKINSON, R. 1988: 'Ideology and domination' in Aboriginal Australia: a Western Desert test case. In Ingold, T., Riches, D. & Woodburn, J. (eds.), *Hunters and Gatherers II: Property, Power and Ritual* (Oxford, Berg).

TOOBY, J. & COSMIDES, L. 1992: The psychological foundations of culture. In Barkow, J. H., Cosmides, L. & Tooby, J. (eds.), *The Adapted Mind: Evolutionary Psychology and the Generation of Culture* (Oxford, Oxford University Press).

UNGER-HAMILTON, R. 1989: The Epi-Palaeolithic Southern Levant and the origins of cultivation. *Current Anthropology* 30, 88–103.

DE WAAL, F. 1996: *Good Natured. The Origins of Right and Wrong in Humans and Other Animals* (Cambridge, MA, Harvard University Press).

WATANABE, J.M. & SMUTS, B.B. 1999: Explaining religion without explaining it away: trust, truth and the evolution of cooperation in Roy A. Rapoport's 'The Obvious Aspect of Rituals'. *American Anthropologist* 101, 98–112.

WHITEN, A. & BYRNE, R.W. 1997: *Machiavellian Intelligence II: Extensions and Evaluations* (Cambridge, Cambridge University Press).

WHITEN, A., GOODALL, J., McGREW, W.C., NISHIDA, T., REYNOLDS, V., SUGIYAMA,

Y., TUTIN, C.E.G. WRANGHAM, R.W. & BOESCH, C. 1999: Cultures in chimpanzees. *Nature* 399, 682–5.

YOFFEE, N. 1993: Too many chiefs: (or, safe texts for the '90s). In Yoffee, N. & Sherratt, A. (eds.), *Archaeological Theory: Who Sets the Agenda?* (Cambridge, Cambridge University Press).

Index

Illustrations are denoted by page numbers in *italics*.